THERE'S NO PLACE LIKE
PLACE AND CARE IN AN AGE

T0228085

Geographies of Health

Series Editors
Allison Williams, Associate Professor, School of Geography and Earth
Sciences, McMaster University, Canada
Susan Elliott, Dean of the Faculty of Social Sciences,
McMaster University, Canada

There is growing interest in the geographies of health and a continued interest in what has more traditionally been labeled medical geography. The traditional focus of 'medical geography' on areas such as disease ecology, health service provision and disease mapping (all of which continue to reflect a mainly quantitative approach to inquiry) has evolved to a focus on a broader, theoretically informed epistemology of health geographies in an expanded international reach. As a result, we now find this subdiscipline characterized by a strongly theoretically-informed research agenda, embracing a range of methods (quantitative; qualitative and the integration of the two) of inquiry concerned with questions of: risk; representation and meaning; inequality and power; culture and difference, among others. Health mapping and modeling, has simultaneously been strengthened by the technical advances made in multilevel modeling, advanced spatial analytic methods and GIS, while further engaging in questions related to health inequalities, population health and environmental degradation.

This series publishes superior quality research monographs and edited collections representing contemporary applications in the field; this encompasses original research as well as advances in methods, techniques and theories. The *Geographies of Health* series will capture the interest of a broad body of scholars, within the social sciences, the health sciences and beyond.

Also in the series

Therapeutic Landscapes
Edited by Allison Williams
ISBN 978 0 7546 7099 5

Sense of Place, Health and Quality of Life
Edited by John Eyles and Allison Williams
ISBN 978 0 7546 7332 3

Primary Health Care: People, Practice, Place
Edited by Valorie A. Crooks and Gavin J. Andrews
ISBN 978 0 7546 7247 0

There's No Place Like Home: Place and Care in an Ageing Society

CHRISTINE MILLIGAN
Lancaster University, UK

Routledge
Taylor & Francis Group

LONDON AND NEW YORK

First published 2009 by Ashgate Publishing

2 Park Square, Milton Park, Abingdon, Oxon OX14 4RN
711 Third Avenue, New York, NY 10017, USA

Routledge is an imprint of the Taylor & Francis Group, an informa business

First issued in paperback 2016

British Library Cataloguing in Publication Data
Milligan, Christine, Dr.
    There's no place like home : place and care in an ageing
    society. -- (Geographies of health)
    1. Older people with disabilities--Care--Cross-cultural
    studies. 2. Older people with disabilities--Home care.
    3. Older people with disabilities--Institutional care.
    4. Caregivers--Great Britain--Case studies. 5. Caregivers--
    New Zealand--Case studies. 6. Medical personnel-caregiver
    relationships.
    I. Title II. Series
    362.4'048'0846-dc22

Library of Congress Cataloging-in-Publication Data
Milligan, Christine, Dr.
    There's no place like home : place and care in an ageing society / by
Christine Milligan.
        p. cm. -- (Ashgate's Geographies of health series)
    Includes bibliographical references and index.
    ISBN 978-0-7546-7423-8 1. Social work with older people.
2. Older people--Care. I. Title.
    HV1451.M555 2009
    362.6--dc22

    2009020355

ISBN 978-0-7546-7423-8 (hbk)
ISBN 978-1-138-26006-1 (pbk)

Transfered to Digital Printing in 2014

Contents

List of Figures and Tables

Figure

Tables

Preface

At age 84, my father-in-law was diagnosed with dementia, he was a non-smoker who drank alcohol only sparingly at celebratory and other 'special' social events. The onset of his dementia followed closely on the heels of medical surgery. Until that time he had lead a fit and active life and would often pass the time doing woodwork in his garden shed or helping out a friend or neighbour. On more than one occasion he could be found chopping wood for the 'old chap' who lived further along the road he lived in. My mother-in-law was 82 at the time; she had (and still has) partial hearing and a long-standing and incurable sight impairment. Despite these physical impairments, she was a fit and active woman who undertook the main caring role during the two years of my father-in-law's dementia – all but the last six months of which took place in their own home. She was regularly supported in this caring role by two of her non-resident adult children until my father-in-law's health deteriorated to the extent that a 24 hour nursing home care became the only realistic solution. My husband, despite being a non-resident carer, took a significant role in caring for his father, fulfilling many of the intimate caring and personal hygiene tasks that more commonly fall to female family members.

Support from the formal care services at that time can, at best, be described as disorganised and chaotic. Some services were put in place whether required or not and others that were needed were not available. Despite numerous phone calls to cancel deliveries of incontinence pads, for example, they continued to arrive on a weekly basis. At one point, the spare bedroom in my mother-in-law's home was piled from floor to ceiling with boxes of these pads. While incontinence is no laughing matter, my father-in-law was at least guaranteed to remain clean and dry! While it is possible to see some humour in this situation, my later research revealed that in other areas, those who desperately needed these resources had to do without, creating distress and humiliation for all concerned. 'Tuck in' services aimed at helping my mother-in-law to get her husband ready and lifted into bed arrived at 7.00 p.m. at night leaving her with the unenviable task of trying to keep her husband from trying to climb back out of bed when no help was readily available to get him back into it. New paid care workers would arrive at the house with absolutely no formal training in either basic hygiene or care techniques. In one instance a care worker was found attempting to empty the contents of the commode down the kitchen sink. We began keeping a diary of events and in one month, we noted that over seventeen different paid care workers arrived on my mother-in-law's doorstep. Her confusion became so extreme, that on one occasion she showed a (rather confused) woman collecting for charity into my father-in-law's downstairs bedroom, thinking she was yet another in a long line of care workers. Her distress about the quality of care services lead us to contact an NHS arbitrator, whose role was to negotiate with these services in an attempt to bring order to the apparent chaos of care delivery.

This, of course, is an extreme example and my mother-in-law was not without family support. Yet, my later research revealed that such experiences are not uncommon.

Two years after the death of my father-in-law, my widowed mother was diagnosed with Alzheimer's disease. I have lived proximate to my mother for nearly 30 years and with two daughters, the primary caring role fell to both myself and my sister. My mother suffered from an accelerated form of Alzheimer's and died just over a year from diagnosis. The restless and aggressive nature of the early stage of her illness lead to several extremely concerning incidents in which she disappeared from her home for short periods of time, leaving my sister and myself frantic with worry. On one occasion she walked onto a train and turned up in a city some 70 miles away with no recollection of how she got there, or how she could get home. Our concern was such that we took clinical advice and agreed to place my mother in the dementia wing of a psychiatric hospital for a six-week period of assessment. I can only describe this place as truly awful. Whilst architecturally the setting was wonderful, the staff attended only to the personal and medical needs of the patients, there was no attempt at therapy or occupational activity. Those in for assessment were placed in a locked (and mixed) ward, with no locks on the bedroom doors. The restless nature of the dementia suffered by some of those in the ward meant that they regularly wandered in and out of their fellow patients rooms, opening doors and cupboards and taking their belongings. Staff in the unit appeared to have little comprehension of the extent of distress this caused both patients and their families. On one occasion my mother was attacked by a fellow patient, on another occasion, a fellow (male) patient lay screaming across the doorway to her bedroom. My mother refused to go to the communal bathroom in the night for fear she would meet one of these patients. She was continent on entry to the hospital ward and almost fully incontinent on leaving the place six weeks later. Don't get me wrong, I am sure that by the very nature of my mother's illness she is likely to have been responsible for a share of the disruptive behaviour in the ward. But I remain convinced that much of the distress could have been avoided had the hospital ensured that in-patients were able to enjoy a modicum of privacy and that some activities or occupational therapy were available. It was clear at the end of her period of assessment that my mother's mental health had deteriorated to such an extent that we would be unable to care for her at home.

Following much distress and soul-searching my sister and I took the advice of the medical and social work staff assigned to my mother's case and placed her care in the hands of staff in a residential care home specialising in dementia.

The care home we 'chose' for my mother had many good things to recommend it. It was relatively small in comparison to some of the 'chain' care homes in the area, its staff-to-patient-ratio was much lower than that of many other nursing homes and many of the paid care staff were warm, caring people. But we experienced many frustrations as well. Claims were made about activities and treatment prior to my mother's entry to the care home that were never delivered in practice. Privacy, though considerably better than the assessment ward in the psychiatric hospital, was still limited. Conflict and confusion arose over decision-making and medication. We of course are not unique, many others share similar experiences, but all of this is extremely wearing on informal carers. Contrary to the expectation that placement in a care home would relieve my sister and I of some of the stresses of caring, we, in fact, found ourselves facing new and unexpected stressful experiences.

You might be forgiven for thinking that the experiences recounted above occurred around the 1950s or 1960s – after all, it was the work of Goffman and Townsend that first drew our attention to the dire circumstances experienced by many of our disturbed and frail older populations. In fact, the above account of my mother's experience occurred between January 2000 and February 2001 – my father-in-law's experience occurred only two years earlier.

Yet these experiences have not all been bad – I have the highest regard for the social worker assigned to my mother's case – she steered us through the confusing maze surrounding care services in the UK, her support was unfailing during periods of bewilderment and disillusionment in relation to my mother's residential care. Many of the care workers in the care home were wonderfully kind and caring people, as was one of the clinical ward managers. In my view, the remuneration they receive for this work is poor in comparison to the often stressful and difficult tasks they undertake.

Nevertheless, it was the experiences of care during the periods of both my mother's and father-in-law's illness that drew my attention to the apparent disorder and disparities in the system. In the case of my father-in-law, I was more of an onlooker – giving practical support where needed, helping to complete forms and dealing with the formal care services, but with two sons and a daughter living in the same town, my mother-in-law was one of the 'lucky' carers, in that she did not lack for close familial support. In the case of my mother, I experienced these things first-hand. More than seven years on from my mother's death, I still question a system whose legislation pays homage to human rights and dignity, but whose practical application can at times leave much to be desired. Were we unlucky in the services we received or were our experiences 'the norm'? Carers are often so physically and emotionally drained, that summoning the additional energy required to protest about unsatisfactory services may be beyond them. How are informal carers included (or excluded) in decisions made about around care services in the home and in residential care? Where does power and control really lie? How can examples of good practice be identified and replicated? These are all questions that underpin the work on which this book is based.

Christine Milligan
March 2009

For Norman

Chapter 1

Introduction

Care for frail older people often involves not just one story or narrative, but several, each evolving over time and space.

Despite the increased interest in informal care in recent decades, the concept itself is not new. The UK for example, as with many advanced capitalist countries, has long had a mixed economy of care in which the state, the family, the voluntary sector and the market are seen to play different roles at different times. Indeed, Offer (1999) went so far as to claim that the 'classic welfare state' (as epitomised by Britain between 1945 and 1976) should be seen as exceptional rather than a culmination of earlier ideas and practices. Rather than a recent 'discovery', then, the 're-discovery' of informal care by political and academic communities – particularly since the late 1970s – has emerged at a time when idealist conceptions about the nature of 'real welfare' have been discarded.

Three significant factors have contributed to the raising of state and professional interest in informal care, particularly within advanced capitalist societies. First, global ageing – whilst this is a worldwide phenomena, many advanced capitalist states are seeing a particular growth in the 75+ age groups. Indeed, recent figures from the UK Office of National Statistics reveal that for the first time ever the country has more pensioners (defined as those aged 65+) than under-16s (ONS 2008). Whilst many older people do not require care, it is nevertheless true that advancing age increases the likelihood of frailty and hence the potential need for care and support (Williams and Cooper 2008). Second, within many neo-liberalising states there has been a widespread erosion of public sector support services over the last two decades or so in favour of a mixed economy of care. This includes an elevated role for the private and third sectors, underpinned by attempts to reinvigorate 'responsible citizenship'. If we accept Offer's interpretation of the rise of the welfare state, such shifts in care can be seen as a return to the prevailing status quo. Both of the issues referred to above should be interpreted alongside a third shift – an ideological turn towards 'ageing in place'. This shift has taken root in many advanced capitalist states over the latter half of the twentieth century. It has been underpinned by policies and supports designed to enable older people to remain in their own homes as long as possible, with the aim of reducing the need for residential care. In sum these three factors have contributed to an increasing shift in health and welfare policies and practices in neo-liberal states away from state dependency models of provision in favour of personal and family reliance. Inherent within this shift are assumptions not just about the nature of the family in contemporary society, but also its willingness, and capacity, to assume caring responsibilities. These changes are of particular significance for the long-term care of older people in an ageing society,

While policy and legislation in many neoliberalising countries acknowledges that there is still a role for statutory and independent bodies in supporting older people to 'age in place', in reality, where care support in the home is required, most is undertaken by family, friends and neighbours. Such care is largely unpaid and can involve a wide variety of tasks ranging from shopping and the management of financial matters to personal care and medication. Within this book such care is referred to as *informal care* as distinct from *formal care* consisting of paid care work undertaken by a range of health and social care professionals drawn from the statutory, voluntary and private sectors. In the UK, as many as one in eight people are now seen to undertake informal (or family) care, looking after someone who is sick, elderly or disabled (Department of Health 2008) – nearly five million of these people care for someone who is over the age of 65 (Maher and Green 2002). Some five years ago, the UK carers' charity Crossroads estimated that the overall saving to government and taxpayers was around £57 billion per year (*BBC News* 2003). In the United States, it has been estimated that around one in six households (around 35 million people) provide unpaid care for an adult over the age of 50 – mostly to parents or grandparents (National Alliance for Caregiving 2006). Increasing numbers of caring households are also evident in such countries as Australia, Canada and New Zealand. Clearly, then, informal care within the home is a key component of policy around care for older people and, as such, is crucial to its successful implementation.

Few would argue that state provision could ever replace that provided by informal caregivers. Indeed, Hirst (2002) maintained that most community health and social care services would be unable to cope without the contribution of informal carers – a comment echoed by ter Meulen and van der Made (2000 p. 257) who maintained that, 'Informal care not only supplements professional care, but is a basic conditioning for the functioning of the organised health care system'. Commenting on the extent of informal care-giving in the UK revealed by the 2001 Census, the then Liberal Democrat spokesman for older people, Paul Burstow, went as far as to claim that informal carers are the foundation upon which the whole care system stands, without them, the whole system would face collapse and many voluntary organisations would struggle (*BBC News* 2003). Supporting informal carers is thus high on the public policy agendas of many advanced capitalist countries.

However, in conjunction with this recognition that informal carers are critical to the success of policies designed to support 'ageing in place', there has been a growing acknowledgement that the heavy demands of care-giving can lead to a decline in the physical and mental health of informal carers themselves (O'Reilly et al. 2008). One consequence of this has been that in countries such as the UK, Canada and New Zealand, governments have introduced a raft of strategies, guidance and legislation designed to support informal carers. This not only places them at the centre of future health and social care strategies, but in the UK at least, is seen to indicate that informal carers are to be supported and viewed as *partners* in the care of their family member. Indeed, the 2001 National Service Framework for Older People explicitly stated this, thus promoting a co-worker model of formal and informal care.

'Ageing in place', then, not only elevates the domestic home as the preferred site of care, but it increases the complexity of the relationships between formal and informal care-givers and care-recipients within the home. Care-giving for older people involves a complex pattern of people (formal and informal carers), places (domestic/non-domestic) and times (daily, intermittent, continuous). As such, it involves not just a single, situated narrative, but several, each evolving over time and space. These narratives do not occur in discrete spaces; rather their complex dimensions are manifest through a series of interwoven stories that emerge at different times as a result of differing sets of circumstances. They can stretch across and beyond the domestic home, to include the community, public and private institutional settings. Understanding how older people and their informal carers experience these shifting landscapes of care is important, but as yet, we have little understanding of how these complex stories weave together. Further, the explicitly spatial nature of care-giving is relatively under-researched and while health geographers have begun to engage with these issues, to date, this work has been poorly conceptualised. *There's No Place Like Home: Place and Care in an Ageing Society* represents one attempt to redress these gaps by focusing on those relational practices of care-giving to frail older people that extend from the domestic space of the home to the community and institutional care homes. The book, thus, engages with critical concerns about the nature and site of contemporary care and care-giving. Its geographical lens seeks to provide an increased understanding of how the increasingly porous boundaries between formal care-giving and informal care create complex landscapes and organisational spatialities of care.

Unpacking these complex spatialities involves an understanding of the politics of welfare, prevailing ideologies and socio-cultural systems. Such an approach can enhance our understanding of the needs and interests of dependent groups and their care-givers as well as issues surrounding the social and spatial equity of care.

It is against this background of the growing importance of informal care-giving within academic and political discourse, but also more critical concerns about the nature and site of care and care-giving, that this book is located. It focuses firstly, on the spatial manifestation of care at various scales, and how care is woven into the fabric of particular social spaces, identifying some of the processes behind variations in the care-giving experience. In particular, attention is drawn to the need for a greater understanding of how the interplay between local cultural practices, the social politics associated with care provision and wider structural forces impact on care-giving behaviour within and across space. Secondly, the book engages with debates around the meaning of care and how space can be conceptualised within these debates. More specifically, the book focuses on how the intersection of formal and informal care-giving within domestic, community and residential care settings create a particular landscape of care for older people. To what extent, for example, might it be said that increased dependency on formal care services within the home contributes to an institutionalisation of the home for some frail older people (Milligan 2000)? The growing complexities surrounding these landscapes of care are further called into question in considering care transitions from the domestic home to supported accommodation and residential care homes. For while there has been a burgeoning of interest in the changing nature of care and what this means for informal carer givers (particularly within the sociological and social policy literature) this work has centred almost exclusively on community-based settings and is aspatial in its analysis. Limited attention has focused on how informal care-givers experience and cope with immediate and longer-term transitions in the place of care. It is generally assumed, for example, that on entry to residential care, the 'burden' of caring transfers from the informal carer/s to professional caring staff within the care home setting. Yet, informal carers can undertake a surprisingly high level of care in care homes, highlighting the porosity of boundaries between formal and informal care. These transitions and how they impact on the lives of informal carers and care-recipients underpin the analysis of care in this volume.

Not only has the spatial nature of care-giving for frail older people received limited attention, but perhaps with the exception of Twigg's (2000) work on the care in the home, it has also been poorly theorised. As she, herself, noted, while such care touches on some of the most intimate and important aspects of people's lives, it has been dominated by practical concerns and as such, has tended to become something of an academic backwater. The book, thus aims to conceptualise the spatial experience of care in terms of informal carers' location and dis-location across the landscape of care. Here, I draw on ideas about anthropological place and non-space postulated by Augé (1995); Serres (1995) notion of extitutional arrangements; and the institutionalisation of the home (Milligan 2000) as a means of interpreting the shifting relationships between people and places in the construction and delivery of care to older people. Running through this book, then, is an engagement with debates around the place of formal and informal care-giving for older people and what this might mean in terms of inclusionary and exclusionary sites of care.

The book draws on debates within the published literature, together a range of case study material and secondary data (largely but not exclusively) from advanced capitalist countries to frame a critical discussion of care-giving for older people across both community and residential care settings. It does so in a way that contributes not only to our appreciation of the importance of place in the care-giving relationship, but also to our understanding of life-course transitions experienced within the care-giving relationship. As articulated through the everyday life experiences of older people and their carers, the book gives attention to the material and non-material manifestations of care within the home, community and residential settings. The aim here is to draw together and extend ideas and understandings emerging from research that I have undertaken with older people and their informal carers – mainly in the UK but also in New Zealand – over the course of the past ten years. Whilst individual studies have had a different overall focus, all have been concerned to explore the relationship between people, place and care. More specifically, they have sought to draw out issues of dependence and independence, need and access to formal and voluntary care services as well as the experience and implications of care for the lives of research participants.

Though much of the focus of this book is on the care of older people within the UK, many of the core themes and concepts discussed are of wider relevance. Hence it is hoped that the book will have resonance for those interested in care and older people across a range of western – particularly neo-liberal – settings. Conceptualising the landscape of care, then, is underpinned by the need to draw out (the complex and evolving) manifestation of care in particular places through a particular social science perspective. In doing so, the book seeks to progress understanding and debate along both these fronts, drawing attention to those largely neglected aspects of informal care for older people and their interrelationship with formal care and places that warrant further examination.

Before outlining the structure of the book, it should be noted that in addition to the definition of informal care-giving discussed on page two, there are three further terms used in this book that perhaps warrant some clarification. Firstly, it is acknowledge that there is substantial debate around what constitutes an older person and whether 'old age' should be chronologically defined or linked to some level of physical limitation or disabling condition arising as a result of the ageing process. This is discussed in more detail in Chapter 4. In most western settings old age is linked to pensionable (hence chronological) age – most commonly between 60-65 years. As that is the age around which most statistical data on older people is collated and analysed, this is the definition adopted within this book. When referring to 'frail elderly' however, the book is concerned with older people who experience a level of physical limitation or other disabling condition that requires them to seek care or support in undertaking activities of daily life. Secondly, the term *supported accommodation* is used to refer to (usually) specialist housing options for older people that offer on-site support. This may range from an on-site warden and housing with built in care-call systems and home adaptations, to 'Extra Care' with a wider range of support workers and more intensive '24/7' support. Such accommodation may be referred to elsewhere as assisted living, Sheltered Housing or similar. Thirdly, the term *care home* is also used here to refer to long-term residential settings that provide either, or both, nursing and personal care to frail older people who are no longer able to live independently at home and/or where community-based care support has broken down. This reflects the different terms used in different national settings and the varying stages at which people enter specific types of residential care.

Though each chapter is constructed as a complete entity, the book is designed to provide an ongoing narrative about care and transition and the impact on older people and their informal carers. Chapter 2 situates the volume relative to its broader academic and socio-political context and introduces the notion of 'the landscape of care'. It suggests that the landscape of care is a multi-faceted concept in at least two distinct senses: a) it refers to the *complex embodied and organisational spatialities* that emerge from the intersection of formal and informal care-giving in both domestic and institutional environments; b) the landscape of care can be thought of as a body of intellectual work – one that takes as its starting point the emplaced nature of care and care practices. The chapter begins by giving an overview of the growing debate around informal care and care-giving in contemporary western society. It critically reviews how political and ideological shifts in welfare are impacting on the meaning and place of care in the late twentieth/early twenty-first century. This approach sets the scene for debate about the changing nature of the care-giving relationship and its spatial expression. The chapter also critically reviews some of the key theoretical approaches that social scientists have used to explore the shifting topology of care.

Chapter 3 draws on recent case study and statistical data in neo-liberal states to examine who actually undertakes the informal care-giving role and where. In particular, it explores issues of gender, class and ethnicity. These issues are framed within contemporary conceptual and theoretical debates about: *proximity* and *distance* the extent to which the changing nature of the family may be impacting on how informal care is conceptualised within co-resident and non-resident frameworks; debates around the *gendered* nature of care and the extent to which this may be manifest in an institutionalised gender bias in available care support. The chapter also considers the less frequently examined issue of care-giving and *socio-economic status*. That is, the extent to which people's willingness and ability to care may be wrapped up with issues of social class and cultural norms. It also briefly addresses issues of culture, ethnicity and care. In sum, the chapter considers the extent to which social and spatial variations in contemporary society affect *who cares*.

Whilst the book is framed largely around informal care-giving for older people within western democracies, as sugessted above, definitions of what constitutes care and the experience of informal care-giving are highly cultural (Tronto 1993; McDaid and Sassi 2001). The availability of formal and informal care across space and between differing social groups is, thus, subject to varying social and political perceptions of rights and responsibilities in the field of care-giving. Chapter 4 thus serves to give an overview of cultural and international differences in the meaning and practice of informal care-giving. Drawing primarily on secondary data sources, it considers how ageing and informal care-giving is understood and experienced within differing national and international contexts. In doing so, it draws out some of the common, as well as some of the culturally specific factors, that have contributed to shifts in who cares where for older people in differing settings across the globe.

Chapter 5 focuses on care and the home. More specifically it addresses the role of formal and informal carers in supporting frail older people to age in place. Framed by the contemporary legislative and policy context of care for frail older people and informal carers in the UK, it examines how both carers and care-recipients experience care-giving in the home. In doing so, it draws out debate around the home/care dichotomy where the complexity of the home as a site of care, work and personal meaning brings both public and private into tension.

The current and projected growth of those in the older age groups (particularly amongst the oldest old), shifting family structures, the potential 'care gap' combined with policies designed to support 'ageing in place' raises increasing dilemmas about how this care will be actualised. One response has been a turn to new care technologies as a means of supporting the care needs (or perceived care needs) of frail older people within the home. This 'technological fix' raises some important questions about how older people and their care-givers experience these new care technologies and the extent to which they may be reshaping both the place and nature of care. New care technologies, for example, create shifts in care work and responsibilities to care. In Chapter 6, then, I consider how these new technologies may be contributing not only to a reshaping of who undertakes care and the nature of that care they perform, but where that care takes place. To what extent, for example, do new care technologies hold the potential to make the homes of older people 'better places to live in' – or do they act to increase social isolation and change the spaces and functions of the home such that the home is no longer a recognisable or desirable places to be?

Care within the domestic home does not exist in a vacuum; hence Chapter 7 is concerned with issues of community, care and support outside the domestic (or familial) home. It considers how informal carers understand and gain access to public, private and voluntary care services and how place can play an important role in mediating the availability and access to these services. Whilst ageing in place is often equated with staying put within the domestic home, we have also seen the emergence of a range of alternative supported housing options. Hence this chapter also addresses issues around the experience of transition to supported housing. Finally, the chapter engages with some of the critical debate around care and community that suggests that globalisation and increased mobility over the life course may be limiting the relevance of community and place for older people.

Chapter 8 addresses care and the transition from community to residential settings. Many informal carers undertake a significant proportion of the caring role in the domestic home; hence the care transition can result in growing confusion over their rights and responsibilities within these new sites of care. Informal carers can thus find themselves having to construct new roles and identities for themselves. For some, this may simply be a transition from primary carer to 'visitor', but many others continue to play an active role in care-giving to their spouse or close family member. The level of informal contribution to the care of older people in care homes can be surprisingly high (Belgrave and Brown 1997; Milligan 2006). Indeed, though informal care given within the domestic home as been referred to as 'invisible care' that rendered within care home settings may be even *more* invisible. This chapter, then, is not just concerned with the transition of care, but also examines the forms and extent of care work undertaken by informal carers in care homes and how they negotiate new identities for themselves within these new care settings.

In Chapter 9, I discuss emotion and the socio-spatial mediation of care. Firstly the chapter draws attention to the complexity of factors that contribute to variations in the health and emotional well-being of those involved in the care of older people. In part, this stems from literature on 'expressed emotion' drawn from the field of clinical psychology (e.g. LeTreweek et al. 1996); in part, from health research that has focused on emotion and care-giver 'burden' (e.g. Burns 2000). The chapter then goes on to discuss the small, but growing body of literature around emotional geographies of care that addresses both the embodied emotional experience of informal caring and the affective entity of informal care work. Drawing on empirical case material the chapter also discusses how emotion is both experienced and performed by informal carers in a range of care settings.

As the narratives of care discussed in the previous chapters indicate, care does not occur in discrete spaces, but stretch across and beyond the domestic home, to include the community, public and private institutional settings. In Chapter 10 I thus consider how the narratives of care outlined in the previous chapters are manifest in an increased porosity between the worlds of formal and informal care and the places in which that care occurs. Increased porosity suggests the emergence of more inclusive sites of care – but how is this experienced in practice? Drawing on case material, this chapter goes on to discuss what contributes to the development on inclusionary and exclusionary landscapes of care. This in turn gives rise to further discussion about how care can be conceptualised within debates around institution and extitution.

This penultimate chapter is followed by a brief concluding commentary that draws together and summarises the differing dimensions of care explored within the volume. It highlights the ways in which the theoretical and empirical analyses offered here contribute to a new geographical analysis of care; a perspective largely unexplored by those researching the informal care-giving through an historical, gerontological or social policy lens. Finally it reflects upon some notable gaps in the landscapes of care literature suggesting a number of potential directions for future research into the place of care.

Finally I would like to acknowledge that whilst the book is concerned with landscapes of care for older people, my key concern has been with how care is experienced within and across home, 'Home' and community. The very complexity of these landscapes means that inevitably there are aspects of these spatialities that I do not address within this text. In particular the urban–rural dimension deserves further consideration as do more detailed trans-national comparisons. Nor do I attempt to address the issue of child-carers – and important area in developed, but particularly developing, countries. I leave these and other unaddressed aspects of the landscapes of care to the expertise of others working in this field.

Chapter 2
Conceptualising
the Complex Landscapes of Care

This chapter begins by giving an overview of the growing debate around informal care and care-giving in contemporary western society. It discusses political and ideological shifts in the meaning and place of care in the late twentieth/early twenty-first century and the implications for contemporary care-giving for frail older people. The chapter also critically reviews some of the key theoretical approaches that social scientists have used to explore the shifting place of care within society and within welfare state reform. In doing so, it suggests that the landscape of care needs to be understood as a multi-faceted concept in at least two distinct senses: firstly, it refers to the complex landscapes and organisational spatialities that emerge from the intersection of formal and informal care-giving in both domestic and institutional environments; and secondly, it represents a body of intellectual work. Within the discipline of geography, for example, there is now an identifiable strand of research that focuses specifically on how and where care takes place and how it is experienced by frail older people in our society and those involved in their care. This ranges across a spectrum from work on the shift from institutional to deinstitutionalised care, including care within the home and community-based settings (e.g. Twigg 2000; Milligan 2000 2001; Williams 2002; Brown 2003 2004; Conradson 2003; Wiles 2003 2005; Parr 2003; Skinner and Rosenberg 2006); to how care is conceived of, and delivered, within residential care settings (e.g. Willcocks et al. 1987; Rowles and High 1996; Peace and Holland 2001; Peace et al. 2006; Valins 2006). While individual studies located within these settings have employed differing analytic frameworks, they share in common recognition of the emplaced nature of the social practices and institutions of care.

Conceptualising Care

How care is constructed and the implications for the caring relationship have been the subject of debate within sociological, social policy and feminist research for well over 20 years (e.g. Graham 1983 1991; Twigg 1989; Ungerson 1990; Tronto 1993; Davies 1995; Thomas 2007). Work in this field has been important in contributing not only to policy, feminist, disability and gerontological debates about the nature of the caring role, but also in illuminating the material and affective aspects of the care-giving experience. Feminist approaches in particular, have demonstrated how care has been socially constructed as a gendered concept

– one that reflects a specifically 'feminine' expression of society that is built around notions of familial ties and obligations (see Chapter 3). Such approaches illuminate two analytically distinct, but inseparable, dimensions of care: caring as a material entity (the physical labour of caring); and caring as a psychological or emotional entity (the affective aspects of caring). Some two decades ago Graham (1983) maintained that the emotional investment in the caring relationship marked it out as a private activity – hence one that exclusively located it within the private sphere of the home. Only when care within the home breaks down is it seen to enter the public or market sphere.

These analyses are useful in that they point toward the spatiality of the care-giving relationship but they also have their limitations. 'Caring *for*' as a material entity is often viewed as occurring in specific places at specific times, whilst the affective state of 'caring *about*' is seen to be less bounded by place and time. Indeed, a whole raft of geographical work emerging (at least) since the 1990s has focused on care and responsibility within a global framework. Drawing on debates within moral philosophy, this body of work has been important in highlighting issues of responsibility and social justice as well as some of the ethical and moral dilemmas in caring *about* distant others (e.g. Harvey 1996; Procter and Smith 1999; Smith 1998 2000 2005; Silk 2004). But as Lawson (2007) points out, the issue of proximity and distance in relation to care is more complex than such debates would suggest. The globalisation of care, migration and shifting family and work patterns, combined with the rise of new communication and travel technologies are shifting both the ways and places in which people engage in care. Despite being geographically distant, families can engage in both the affective (caring *about*) and physical (caring *for*) performance of care. That is, that distance carers can perform material care through the organisation of more physically proximate care and the transfer of remittances for the purchase of that care; by monitoring care through modern communication and care technologies; or through engaging in regular travel to perform aspects of that care themselves. At the same time, it is important to recognise that geographical proximity and the performance of the material entity of care is no guarantee of caring *about*. That is, an informal carer may be geographically proximate and engaged in the performance of caring *for*, but emotionally distant. Rather than being marked out as a private activity located solely within the domestic space of the home, then, the material and affective aspects of caring for and about, can occur across both public and private spaces that stretch from the highly place specific to the global.

Whilst it is important to draw attention to debates around care that transcend the national and local, this is not the primary focus of this book (though see Chapter 4). Rather the focus is on how care is manifest within and across the domestic home, the community and institutional spaces. It cannot, for example, be assumed that once a care-recipient has entered long-term residential care that the informal care-giver will dissociate from the physical work of caring (Milligan 2005). Neither can it be assumed that the affective aspects of care are absent from formal care-giving whether that care takes place in the domestic home, the community or within institutional settings (such as long-stay wards, residential care homes and so forth). Caring relationships in these settings can, and often do, involve varying degrees of emotional attachment. Formal carers can develop close emotional bonds with those for whom they provide care and support. So while the affective nature of this caring relationship will differ from that of the informal carer, in that it lacks a shared history and identity, it cannot be assumed that formal care-giving lacks emotional investment. Indeed, a recent review of research around palliative and hospice care clearly illustrates that the delivery of emotional care and support to patients and their families is a critical component of formal care work in this field (Skilbeck and Payne 2003). Furthermore, the giving of this emotional labour is not without cost. The landscape of care, then, can be seen to refer not just to those geographical settings within and across which care takes place, but to a subjectively experienced phenomenon.

We thus need to question any suggestion that the worlds of formal and informal care are discrete and operate within different spheres of knowledge, action and understanding. This requires us to rethink the distinctions between those formal and informal care-giving relationships that develop within domestic, community and institutional spaces. Indeed, adopting any such polarised view will almost inevitably lead to a fragmented understanding of care and the caring relationship, hindering the development of a wider understanding of the social and emotional divisions of care that are performed within and across differing spaces. The giving of care may thus involve either, or both, formal or informal care work performed within a diversity of those social relationships of production that occur within private, public and domestic space as well as reproducing people of differing gendered and socio-cultural backgrounds.

State welfare and shifting typologies of care

How care is conceptualised and performed is also underpinned by differing ideologies of care and how these are played out through state welfare regimes. As Daly and Lewis (2000) point out, care is a growing concern for welfare states, in part due to the growing demand for care; and in part to changing norms about family and kin responsibilities (particularly female responsibilities) that are decreasing the supply of care at a time when demand is rising. Indeed, they suggest that 'it is impossible to understand the nature and form of contemporary welfare states without a concept like care' (282).

At the same time, who cares is also linked to how care is constructed within different forms of state welfare. In his highly influential book *The Three Worlds of Welfare Capitalism* (1990), Esping-Anderson pointed to the existence of three main regimes of welfare that predominate in advanced capitalist states:

Firstly he points to *social democratic regimes* based on universal inclusion and a comprehensive definition of social entitlements. Typified by the Nordic countries of Denmark and Sweden, such regimes have had a commitment to equalising living conditions across their citizenry. Here, private welfare markets and targeted (residual) welfare have been marginalised in favour of a service-intensive welfare state in which the standards and quality of welfare distributed are identical for both rich and poor. The focus is one of de-familialising welfare responsibilities, especially care to children and older people, so freeing up women to become an integral part of the labour force. The model of social service provision amongst these regimes, though changing, has been manifest in abundant, locally-organised services that are available on a universal basis and funded from taxes (Daly and Lewis 2000).

Secondly, he identifies *conservative regimes* whose foundations lie in social insurance largely premised on occupational distinctions. Entitlements are thus dependent on lifelong employment and ensuing pensions. Typified by much of continental Europe, such regimes are strongly familialistic, assuming that the primary welfare responsibilities lie with the family. The expectation is that by and large, women will take on the responsibility for care. Such regimes thus tend to reinforce the male breadwinner model. Private welfare is marginal, in part due to the generous benefits accrued through social insurance and in part due to the high cost of private welfare.

Thirdly, he points to *liberal regimes* that favour minimal public intervention on the assumption that the majority of citizens can obtain adequate welfare through the market. The state's role is thus one of nurturing, not replacing, market transactions. Typified by the United States, public welfare is minimal and targeted at the most needy – generally through some form of means-testing. Care is thus the responsibility of the family, who is expected to purchase support services, respite and so forth through the market.

More recently (particularly from the late 1970s onwards) we have seen the emergence of *neo-liberal regimes*. Here, countries such as the UK, New Zealand, Canada (but also the United States) have sought to roll back their state welfare regimes in favour of the promotion of a mixed economy of welfare. The underlying rationale for such a shift is threefold: firstly that the welfare state, itself, has come under increasing pressure; secondly, debate has emerged around what the 'proper scope' of action by the state in relation to individuals and their families should be; and thirdly, governments have adopted the view that there should be greater consumer participation in the provision of services (Esping-Anderson et al. 2002). Neoliberalising regimes have thus sought to elevate the role of the citizen consumer, emphasising not just individual rights, but also citizen responsibilities for their own health and welfare. Such regimes seek to encourage partnership working and the provision of welfare through a mix of market, state and third sectors.

Differing regimes of welfare and changes in state welfare have implications for how and by whom care is performed. Researchers have sought to encapsulate this within a framework of care. The threefold typology first developed by Twigg (1989) and later refined by Twigg and Atkin (1995) for example, is now widely recognised. Whilst one of the limitations of this typology is that it takes as its foundation the perspective of care professionals rather than informal carer-givers or care-recipients, it does provide a frame of reference through which to examine the different forms of relationship that exist between formal and informal caregivers (Ward-Griffin and McKeever 2000). Further, whilst acknowledging that no one model prevails and that the distinctions between models are often blurred, they nevertheless provide the basis for a useful typology 'of ideal types':

Carers as resources   Here informal care-giving and the availability of informal care are taken for granted. Normative expectations of care mean that this model is viewed as a morally desirable option by the caring professions. Services are designed solely on the basis of the care-recipient's needs – hence the needs of the informal carer are largely ignored. Within this model there is also very little questioning of family members' suitability to care. Informal care is thus the lynchpin in maintaining an older person's ability to age in place.

Carers as co-workers   Here informal carers are seen to have co-responsibility alongside care professionals for the care of their frail older relative. On the one hand they are often seen by care professionals as needing education in how to care and deal with the care-recipient; on the other there is a greater level of consultation and recognition of the informal carer's expertise. The co-worker ideal draws the informal carer into the orbit of the formal care system to become 'semi-professionalised'. The overall goal is to provide the highest possible quality of care through the breaking down of traditional boundaries between formal and informal care. Professional conflicts thus need to be resolved. However, this model does raise ethical concerns about confidentiality and consent as it requires shared communication about the care-recipient that may be at odds with the care-recipients' desires (see Chapter 6). This model also highlights the porosity of the boundaries between formal and informal care.

In extrapolating from this model Gilliatt et al. (2000) also point to the emergence of what they refer to as a 'do-it yourself' model of co-worker. Here, the informal carer is directed to sources of expert advice with the expectation that this will be sufficient to equip them to take on the role of a semi-professional care-giver. While this model sits comfortably with the concept of the carer as expert, as Manthorpe et al. (2001) point out, it can also act to create a dichotomy between the individual as an informal carer and as a family member, where each is given a different status in public discourse.

Carers as co-clients   Whilst the needs of the care-recipient are generally privileged, under this model informal carers are also viewed as individuals requiring support. This is legitimated through policy and service provision. However, this can also act to create conflict between the needs of the carer and those of the care-recipient. Respite care for example, is designed specifically to support the informal carer, but may be seen as inappropriate or an unwanted service by the care-recipient.

Manthorpe et al. add a fourth dimension to this typology – one they refer to as 'the superseded carer' (477). Here, a conflict of interest arises between the desire of the informal carer to care and their ability to perform that care in a way that serves the best interests of the care-recipient. While some informal carers are vulnerable and require help and support themselves in order to continue caring, others are determined to protect the cared-for, and as such are reluctant to use formal care services. These carers may refuse services in the mistaken belief that they can better provide them themselves, or that the services are inappropriate. This model suggests that formal services may help to support such carers and the care-recipient by starting what, from their perspective, is seen as a healthy process of separation.

Whilst recognising that these dimensions represent 'ideal types' and that elements within this typology may be not be discrete in practice, it is nevertheless possible to illustrate how they relate to differing welfare regimes (see Table 2.1). The implications in terms of how care is conceptualised is outlined in Table 2.2.

Typologising the care relationship

Ageing in place means that the vast majority of older people requiring care and support now remain within existing family relationships and settings. Informal carers and care-recipients are actively involved in the creation and maintenance of the form and meaning of their lives – not least through their interaction with each other and others involved in the care of the older person. As with the old institutional context, informal carers are required to accomplish the delivery of that care within a restricted space (the home and immediate surrounds). If care is likely to be sustained over a long time, not only is that space likely to shrink as the intensity of care required increases, but the informal carer will need to conserve his or her efforts in order to be able to care longer.

Any attempt to typologise the care relationship between informal carers and care-recipients thus needs to recognise that this relationship can exist in different forms. Askham et al. (2007) point to a useful threefold typology of custodial, homelife and intimate relationships.

Custodial relationships Characterised by routinisation, surveillance and 'mortification of the self' (4) – the latter refers to the breaking down of the care-recipient's former concept of self and a rebuilding – in part through a process of punishment and reward. Work by Lee-Treweek (1996), for example, has highlighted the way in which some formal care workers seek to control potentially difficult care-recipients and make their own work life easier by 'rewarding' and 'punishing' care-recipients through the giving and withholding of affection. Mortification can also arise through constructed dependency, manifest in such actions as objectification and infantilisation of the care-recipient, invasion of their privacy, talking *about* them rather than *to* them, and responding on the care-recipient's behalf.

Table 2.1   Construction of informal care within differing welfare regimes

Welfare regime →	Liberal	Neoliberal	Conservative	Social democratic
Healthcare system	All healthcare privately paid for. Seen to maximise choice for those able to pay. State plays residualist role with only very poorest eligible for minimal state-funded healthcare.	Risk managed through managerialist doctrine. Welfare recipients seen as individuals or consumers. State plays supportive but increasingly residual role.	All healthcare funded through social insurance based on occupational distinctions. Private welfare marginal.	Promotion of collective social conscience. Healthcare freely available to all at point of entry. Focus on meeting needs.
Funding	Individual – through self-funding or insurance schemes.	Mixed economy of welfare including services funded through state taxation, self-funded private healthcare and voluntary welfare.	Social insurance through taxation.	Funded through state taxation based on ability to pay.
Rights and responsibilities	Citizens seen as responsible for their own welfare – promotion of moral obligation and obedience.	Citizens seen to have both rights and responsibilities.	Citizen entitlements to welfare based on life-long employment.	Promotes citizen entitlements and rights to welfare rather than responsibilities.
Active or passive welfare	Promotes consumerism and individualism – people seen as largely indifferent to needs of others and committed largely to personal well-being.	Promotes active welfare through encouraging entrepeneurialism and prudentialism.	Promotes family dependency and citizen responsibility through active welfare.	Promotes clientelism and dependency model of welfare characterised by 'nanny' State and passive welfare.
Implications for informal care	Informal carers seen as resource and the given against which state services are structured, hence carers marginalised as subjects of formal care support. State plays residualist role re support to care-recipient.	Informal carers as co-workers or co-clients. Formal and informal care intermeshed. Interventions aimed at relieving carer stress. Assumes informal carers want to care hence support aimed at maintaining their ability to do so.	Informal carers seen as primary resource. Are the given against which state services are structured. State plays residualist role only where family care absent. Informal carers limited subjects of formal support.	De-familialisation of care-giving, hence informal carers not seen as a subjects for care support.

Table 2.2   Unpacking the concept of care

	Macro-level	**Micro-level**
Conceptual reference →	Division of care (labour, responsibility and cost) for frail older people between the state, market, third sector, family and community.	The distribution of care (labour, cost and responsibility) among individuals within the family and community and the character of state support for caring and carers.
Empirically indicated by →	- The care infrastructure (policies, services and cash). - The distribution of provision within and between sectors.	- Who performs the caring. - Who is the recipient of available benefits and services. - Character of relations that exist between the caregiver and recipient. - The economic, social and normative conditions in which caring carried out. - Economic activity patterns of women of caring age.
Trajectories of change →	More/less provision by state, market, third sector, family, community.	- An alteration in the distribution of caring activity. - An alteration in the identity of carers. - An alteration in the conditions under which caring is carried out and the nature of the state's role therein. - An alteration in the relations between care-giver and receiver.

Source: Adapted from Daly and Lewis (2000, p. 287).

Whilst informal carers often reveal some of these traits, most still attempt to retain the identity of the care-recipient through including them in conversations, ensuring their requests, preferences, likes and dislikes are understood and met and that where possible they are given own personal space. Permanent mortification of the self is rare as care-recipients are likely to use various tactics to maintain their self-identity (Askham et al. 2007).

Whilst surveillance is also seen to be inherent within the custodial relationship, it is important to note that is not necessarily a one-way process, informal carers and care-recipients can monitor each other. Care-recipients, for example, may feel a greater sense of security when they know where the informal carer is and what they are doing. Nevertheless, surveillance is more problematic because unlike routine, it is not part of normal home-life or intimate relationships. Not only does it prevent the informal carer from doing other things, but it can result in resentment on the part of care-recipient who will look for ways to disrupt or evade it. Nevertheless, as Askham et al. (2007) point out, without the aid of new care technologies (see Chapter 6) surveillance in the home can be problematic, not least because the domestic home is constructed with a specific form of living in mind

– one that allocates tasks and activities to specific rooms and spaces and separates them out on that basis. Without the aid of technology, such a construction is not conducive to surveillance.

Homelife relationships   Characterised by stability, changelessness and habituation patterned by long-established habits and timetables. They also infer that these relationships occur within a place where an individual can be him- or herself in a controllable environment that is free from surveillance.

Routinisation in the homelife relationship is common and often imposed by the informal carer as it provides a means of coping with what can be a demanding regime of care and the need to try to maintain some personal time. It largely occurs around tasks such as personal hygiene, dressing, meals, medication and so forth. This can be reinforced by the timing and availability of formal care support both within and outside the home (for example, community day-care, in-home respite, help with bathing, specialised transport services etc.). Hence both the informal carer and the care-recipient often see their day as highly structured and routinised. My own research involving the daily diaries of informal carers, for example, highlights the structured and repetitive nature of their lives. In some cases following an initial detailed outline of their day, informal carers simply filled in 'same as yesterday'. Indeed, informal carers have commented on their own dismay at how their diaries revealed the repetitiveness of their daily lives (Milligan 2000). As one informal carer in Askham et al.'s (2007) UK study commented, 'It's the monotony that gets you down … There is no different way. How can you put someone into bed differently? How can you wash anybody differently?' (10). But routines also help the care-recipient to develop a sense of stability and an ability to cope.

Intimate relationships   Characterised by the potential for open conversation on an emotional level between the informal carer and care-recipient that is private from others. This is more common amongst (but not exclusive to) loving spousal or partner relationships. Here, an informal carer may not necessarily identify as a carer, but as a partner, adult child or sibling engaging in a reciprocal relationship. Hence both the carer and care-recipient help each other in the development and maintenance of their personal identities, but also ensure they make time for moments of privacy for reflection. The opportunity for reflection is an important aspect of the intimate relationship in that it allows for shifts and changes in identity.

Whilst this typology provides a useful way of conceptualising the relationship between informal carers and care-recipients, it is important not to oversimplify the separation of these three spheres. Aspects of these relationships will overlap, shift and change as the frailty of the older care-recipient increases and the form of care and support required shifts and changes. The effort required to balance all three can be a difficult one for informal carers. For example, they may not only take responsibility for the care-recipient's hygiene, appearance, safety and so forth, but can also take the view that the care-recipient requires vigilant monitoring and attention. This can result in them adopting some of the patterns of care found in institutional settings (such as surveillance, inflexible routinisation etc.). It is also important to recognise that these three relationships are based on a model of proximate care it which it is largely assumed that the informal carer is co-resident. While distance care relationships may exhibit elements of these 'ideal types', they can be performed in different ways and through different forms of relationships that may involve intermediaries in their actualisation.

Dependence, independence and care

Thus far, the term 'care' has been adopted relatively unproblematically. Whether or not we should even be talking about care and care-giving, however, is the subject of some contestation. Commentators writing within the framework of disability studies, for example, have sought to reject the concept, arguing that it has come to encompass a fairly narrow range of practices entailed in looking after those deemed to be dependent – including frail, old, mentally and physically disabled people (e.g. Young 1990; Morris 2004; Thomas 2007). From this perspective, care is a concept that both creates and reinforces dependency relationships in ways that actively disadvantage particular groups of people in our society. In doing so it gives legitimacy to the exercise of certain forms of authority and control over the lives of those deemed to be 'dependent'. Indeed, in her recent book, Carol Thomas (2007) actively seeks to contest and problematise not just the notion of care, but its inextricable links to need and dependence. Reflecting on different scales of need, she points out that whilst a disabled (or in this instance, a frail older) person may have a greater than average need for assistance in undertaking the daily tasks of everyday life, so too may a busy and 'successful' non-disabled executive who requires assistance in organising and balancing his or her busy work and home life. So whilst one group of people may require support and assistance with tasks such as bathing, dressing, travelling outside the home and so forth, the other may require assistance with the organisation of meetings, travel arrangements, and the smooth running of their home and family life. But whilst the executive is fêted for his or her success and 'independence', the frail or disabled person can find him- or her- self cast as 'dependent', their role in society devalued and they, themselves, sometimes stigmatised. Yet as the following comment from a participant in a recent study around telecare for older people suggests, the distinction between dependence and independence is a fuzzy one – and one that is subject to social construction and reconstruction:

> I have difficulty with this term independence because what you are doing is not making a person independent, but supporting a person in different kinds of ways … is this independence indeed? In our research we also see different forms of dependence appearing because people get more reliant on the healthcare system – they know they are checked, they know that they are cared for. It's a specific kind of independence which is actually very much supported – in this case by technology (Mort et al. 2008, 57).

Whilst this excerpt relates to a very specific example of care, it nevertheless reinforces the point made by Thomas – that in different ways, and to a greater or lesser extent, we are all dependent on others for the smooth running of our daily lives. It is the privileging of some forms of dependency over others that acts to marginalise or privilege an individual with concomitant impacts on the extent of control they are able to exert over their own lives.

Whilst Thomas's work focuses specifically on disabled people, she also makes the point that disabled and frail older people tend to be treated as mutually exclusive categories despite the fact that impairment rises with age and is concentrated in those aged 60 years and over. Yet older people are largely by-passed in a disability literature that tends to focus its attention largely on adults and young people. It would be equally true to say that with the exception of Oldman (2002), social gerontology has, by and large, also failed to engage with disability debates that are, in effect, running parallel to critical gerontology. This failure to engage across the two fields is puzzling, for as both Zarb (1993) and Oliver (1992) have noted, disabled people themselves have complained about the less empowering services they receive once transferred from 'disabled' to 'older people's' services. Furthermore, being identified as both an 'old person' and a 'disabled person' is becoming more common as more disabled people survive longer (Oldman 2002).

Within the gerontological literature, care and care-giving tend to be framed as social problems whether for the family, policy-makers or the long-term care industry (Dannefer et al. 2008). The care-recipient is often viewed from the perspective of passive and objectified helplessness, whilst the care-giver has a position of relative power and authority. Whilst it would be difficult to dispute the asymmetry of the care relationship, the relational nature of care is often stultified in the literature through the production of medical models that are both limited and dyadic in nature. Further this linking of dependency and care varies across places. In the welfare states of continental and southern European, for example, the giving and receiving of care is often viewed as part of the normal reciprocity between individuals who are defined by the nature of their embeddedness in a range of social relations (Daly and Lewis 2000). But reciprocity is not just a feature of care in certain countries, indeed a number of commentators have drawn attention to the need to recognise the reciprocal nature of the care relationship.

Indeed, it is worth noting that the linking of dependency and care is a recent phenomenon. Historically, dependency was not seen in the way it came to be later defined by liberal feminism (Offen 1992). Social exchange theorists have long argued that care-giving – at least within the immediate family – should be seen as part of a relationship of interdependency that has a long history of reciprocal exchange (e.g. from parent to child; partner to partner; adult child to ageing parent etc.) (Call et al. 1999). Constructing some groups as more dependent recipients of care than others can thus obscure the multiple and reciprocal nature of these caring relationships. For older people requiring some level of familial care and support, reciprocity can include babysitting and childminding for grandchildren, the making or baking of treats, financial or other material support, as well as the provision of emotional support. Within this framework, dependency is conceptualised more in relation to the inability of an individual to bring anything of value or exchange to the relationship. Yet as Tanner (2001) points out, whilst those most in need of care and support are likely to have a limited ability to offer instrumental reciprocity, the giving of expressive and emotional support means that reciprocity can be sustained even in fairly adverse circumstances.

It is, of course, important to recognise that some interdependencies will be more balanced than others. It is also true that the balance of power and resources within these reciprocal relationships will shift and change over time. Nevertheless, the balance of care is a critical part of these relationships.

The Shifting Topology of Care

The shifting typology of care and welfare regimes is also manifest in a shifting topology of care. Deinstitutionalisation and the development of community care have increasingly taken root since the mid-1970s. The concept of community care represented an ideological commitment to move away from traditional institutional care arrangements toward the development of new community spaces of care. One consequence of this shift has been a realignment of roles and responsibilities as informal carers, voluntary, statutory and private care providers have been drawn into new sets of care arrangements focused around the home and community. In many advanced capitalist states this shift has been further exacerbated over the last two decades or so by an engagement with neo-liberalist approaches to health and welfare. The concept of deinstitutionalisation has thus underpinned the development of new sets of care practices designed to ensure fewer and shorter in-patient stays. In particular, services are now targeted at facilitating the ability of those with physically and mentally disabling conditions to remain in their own homes for as long as possible.

Much of the early geographical literature around deinstitutionalisation centred on the people/place impacts of shifts from long-stay institutional care to new community-based settings – whether that be the home, hostels, supported accommodation or extra-care settings. Largely focused around those with poor mental health (see for example, Giggs 1973; Dear and Taylor 1982; Dear and Wolch 1987; Moon 1990 2000), this early work pointed to the exclusionary nature of more affluent communities and the consequent agglomeration of deinstitutionalised people in what has been referred to in 'service-dependent ghettos'. Dear and Wolch (1987) for example, pointed to a self-reinforcing process in which deinstitutionalised populations drifted toward more transitory, largely inner-city, locales as they searched for affordable housing, peer and service support. More than 20 years on, the number of people subject to long-term institutional care has declined, with community-based treatment being the norm.[1]

---

1   It is worth noting, however, that some commentators have pointed to a growing 're-institutionalisation' of some individuals with poor mental health within prison settings.

For older people, deinstitutionalisation has been experienced in a rather different form. In the UK, for example, older populations residing in long-term institutional settings were relocated either to private residential care homes, or to psycho-geriatric wings of general or specialist hospitals. Though public sector care homes existed, these tended to be large, often ageing, group homes. The vast majority of these have now be closed down in favour of means tested, state-subsidised places in private care facilities.

Following an initial 'boom' in private care home development in the 1980s and 1990s, we have since seen a gradual closure and levelling off of the number of private care home beds available (see Chapter 8). The assumption is that time and the development of policy and structures designed to facilitate the implementation of care and support within the home, will effectively reduce demand for residential care for all but the most frail older people.

From institutional to extitutional arrangements

Institutional care has been seen to represent a spatial solution to particular 'problems' usually arising as a consequence of frailty and age; mental, physical or intellectual disability. This approach to care has been based on what Vitores (2002, 4) refers to as 'inclusion through exclusion' – the enclosure and 'storage' of particular groups of people away from the public gaze. The physical embodiment of the institution has both internal and external spaces and boundaries, whilst the institution itself is seen to represent the flows of things and people that pass through these spaces, creating routes, habits and situations (Domenech and Tirado 1997). The hospital or residential care home thus represents 'stable stock', whilst the patients and residents represent 'flux' – that is, the flow both into and out of these institutional care settings (Vitores 2002).

The shift to community care and ageing in place, however, has brought into play new sites of care that are remote from traditional institutional settings. The home, day hospitals, lunch clubs, activity centres, call centres and so forth are all examples of sites in which care takes place, but which are remote from both the home and traditional institutional arrangements. But how can these new spaces of care and care arrangements be conceptualised?

It is here that the virtualisation of the institution through the concept of the 'extitution' is of interest (Domenech et al. 2006). The traditional arrangements of attendance based on institutional structures and spaces are replaced, as new emerging entities are identified that may resemble the old institutions, but which are virtual and apart from the building. In other words, the extitution represent a de-territorialisation of the institution and its re-manifestation through new spaces and times that create the potential to end the interior/exterior distinction. This de-territorialisation is characterised by the time-space heterogeneity that embodies the new speeds and spaces through which the extitution operates (Vitores 2002). The following exemplar is drawn from a current ethnographic study that focuses on the impacts of new care technologies on older people and their informal carers:

EXAMPLE: A lone-dwelling older person, whose only informal care comprises distance care from a daughter living some thirty miles away, falls whilst trying to get out of bed triggering her fall alarm. No serious injury is incurred but the older person is unable to get back into bed unaided. The call-centre operator checks the older person's records and assesses the appropriate response. In doing so she judges that the family carer is too far away to respond within the service's target response time; the call-centre operator thus contacts the care services to apprise them of the situation. The care services manager contacts a member of the care response team. The care worker answers the call using her mobile phone and responds to the emergency. The responder is able to gain access to the older person's home through activating the electronic key-safe located outside the front door.

In this scenario, care is taking place within both the physical space of the home and the call-centre but is connected through virtual space by the computer, the phone line and the mobile phone. All form part of an interlinked network of care that is facilitated by the care worker, the care-manager and the call-centre operator. So while the institution has been de-territorialised, it has not been *dematerialised*. In other words, it is still real. Nevertheless, the old institutional way of ensuring a patient or care-recipient remains within the care-giving system based on attendance within a physical (institutional) structure is replaced by a set of associations and horizontal processes (networks) that are dispersed across an open space. As Vitores puts it, 'What matters are the positions, neighbourhoods, proximities, distances, adherences or accumulation relationships' (2002, 4). Each care-recipient's case passes through the centres and institutions involved in the delivery of their community-based care.

The shift from institutional to extitutional arrangements represents a radically different way of inhabiting places. Indeed, Serres (1994) maintains that rather than thinking of such places as being inhabited – a term that infers a period of dwelling over time – we should think in terms of 'haunting' or 'frequenting' across open space. In other words there is no one building to inhabit, rather there is a network of care located in a series of places that are dispersed across open space. The extitution, however, does not consist of any single one of them. Hence we need to thing of community care and ageing in place as apart from the building, that is that older people requiring care and support have no need to go to the institution, the institution comes to them. This takes us beyond traditional ideas about institutionalisation and deinstitutionalisation to think about new spatial arrangement for care that enable us to engage with both the physical and virtual space that these new modes of care inhabit. Indeed, Vitores' (2002) maintains that the conceptual thinking behind deinstititutionalisation is inherently flawed, as in reality it does not represent a shift toward the *extinction* of the institution, but its *virtualisation.* So while the institution, as a concrete entity, may no longer be the stable solution to care for the populations referred to above, it nevertheless *remains* as a way of managing continual change. As the following section suggests, this has significance for how we think about ageing in place.

Reconfiguring the home?

The shift from institution to extitution places greater significance not only on the wider spatial arrangements within which care takes place, but also on the home as a site of care. This raises questions about how extitutional care arrangements may be reconfiguring people's relationship with the home. Geographical research around 'ageing in place' has sought to conceptualise the ways in which it can: a) create changes in how older people and their informal care-givers identify with home; and b) contribute to shifting power relationships within the home, for example, between service recipients, health professionals and family carers (e.g. Twigg 2000; Milligan 2001 2003; Angus et al. 2005). As discussed below, such commentators have pointed to the way in which home-based care places a greater level of power within the hands of the care-recipient, albeit the level of power the individual is able to exert is subject to shifts where dependency on care services increases. Whilst this may be true, care services and technologies designed to enhance the comfort and security of people in their own homes can, in fact, change the nature of home until it is almost unrecognisable. Indeed, as I have suggested elsewhere, the increasing flow of formal care workers and the introduction of even relatively low level technologies such as adaptive aids (e.g. hoists, commodes, lifts etc.) can change the nature and experience of home such that, as frailty increases, an institutionalisation of the home begins to take place (Milligan 2001). The furnishings, décor and layout that are key to how individuals express their individuality and identity, and which have historical meaning for them, are shifted, adapted or removed to make way for care aids and the workspace around them. The boundaries of home and institution become increasingly porous as the private space of the home becomes transformed into a site of work that care-recipients, formal and informal carers inhabit.

This all raises critical questions about what defines good care and whether these new care arrangements make the home a *better* place to live in or whether they change the spaces and functions of the home – and the power relationships within it – such that they may no longer be recognisable or desirable places to live.

The meaning of home in residential care settings

For a significant number of our oldest old increasing levels of frailty and/or a breakdown in the ability of the informal carer to maintain care-giving within the home can result in the transition of care from the domestic home to the residential care home. Whilst a small body of literature has addressed the extent to which residential care can embody the notion of 'home' (Willcocks et al. 1987; Peace and Holland 2001; Milligan 2006) little attempt has been made to theorise its place in the caring relationship.

Though not explicitly concerned with the spatiality of care, Reed-Danahay's (2001) consideration of the borders of home and work for care-recipients and care workers in a residential care setting offers a means of helping us to think through how the transition of care might be conceptualised within a framework that incorporates the changing meaning of home within these settings. Reed-Danahay suggests that in contemporary western society the dementia sufferer can be viewed through the metaphor of displacement – the person who lacks memory and a coherent self-narrative, and who thereby also lacks a proper 'home'. Home is therefore seen to be bound up in individual narratives of meaning. By drawing on de Certeau et al.'s (1998) conceptualisation of everyday inhabited places, however, the metaphor of displacement can be extended to include all those for whom transitions in the place of care represents a disruption to their self-narrative. Hence the portrait depicted by the presence, absence or organisation of objects within the home, and the practices embedded within them, are disrupted by the transition to long-term care. While informal carers and care-recipients are increasingly encouraged to imprint the care-recipient's own identity onto their 'own personal space' (i.e. their bedroom) within the care home, the composition of the self-narrative is disrupted by the order, organisation and preferences of the care home itself. Hence, the organisation of a room may more accurately reflect the 'ways of operating' practiced by formal care workers and that machinery deemed necessary to facilitate their ability to undertake these tasks than the personality or preferences of the care-recipient. Textures and materials used within the care home can also be more reflective of health and safety requirements and the practicalities of caring for a frail older person than any manifestation of the life narrative of the individual inhabiting that space. Communal spaces such as sitting and dining areas offer little opportunity for the care-recipient to imprint anything but a transitory fragment of their life narrative upon them. While such an approach reflects a fairly long-standing geographical concern with the changing meanings of home and self in contemporary society (see for example, Rowles 1987; Tivers 1987; Katz and Monk 1993; Bondi 1998), it is particularly useful within the context of residential care for frail older people as it begins to tease out an understanding of the meaning of place and self for cared-recipients, informal carers and care workers within an environment that can be characterised by changing spatial and social location (or dislocation).

Concluding Comments

This chapter has sought to unpack the concept of care through an exploration of shifting typologies and topologies of care. In doing so, it has drawn attention to the ways in which care for older people is constructed and reconstructed through shifting ideological and political conceptualisations of care, and responsibilities for care, that change over time. These have implications not only for who undertakes that care but where that care takes place. I have also suggested that thinking about emerging arrangements for care through the concept of the extitution may offer greater explanatory power than more commonly used concepts of deinstitutionalisation and community care. Ageing in place and the shift from institutional to extitutional arrangements are also impacting on the worlds of formal and informal care. This is resulting not only in an increased porosity in the boundaries between the realms of the public and private, but also between the worlds of formal and informal care. These are issues that will become increasingly evident in subsequent chapters. Finally, I have also raised the issue of how older people are constructed as dependent and, critically, how care and dependency are often seen as synonymous. Hence, the chapter begins to raise questions about what we mean by 'dependence' and 'independence' and whether the very term care may in fact act to reinforce the construction of older people as dependent.

Chapter 3
Who Cares? People, Place and Gender

The major determinants of the decision to care are based on a range of factors, including a close kinship bond; the nature of the social mores operating on the potential carer; the wider politics of care; as well as the ability to cope. Geographical proximity and distance also plays a part in the form and extent of material care undertaken (Carmichael and Charles 2003). However, the decision to care may also be affected by the availability and cost of alternative sources of support; the financial means of the cared-for and/or the wider family; and the opportunity cost to the potential carer. Yet as already noted, successful ageing in place relies significantly on the availability and willingness of people to undertake this informal care. Ginn and Arber (1992) point out that any policy that seeks to promote familial responsibility for the care of disabled and older people is likely to have a disproportionate effect on women and the working class. As a consequence, the promotion of policies designed to reinforce familial responsibility are likely to compound existing inequalities in informal care. The chapter thus considers these factors in relation to gender and class inequalities in who cares. Whilst the focus is largely, but not exclusively, on the UK, many of the issues discussed will have resonance for other neo-liberalising states. Wider issues of national, international and cultural differences in who cares are discussed in Chapter 4.

Issues of gender, class and care are framed within contemporary debates about: proximity and distance and the extent to which the changing nature of the family may be impacting on how informal care is conceptualised within co-resident and extra-resident frameworks. These debates are also concerned with the extent to which the gendering of care may be manifest in an institutionalised gender bias in available care support. In addressing the less frequently examined issue of care and socio-economic status, the chapter also considers the extent to which people's willingness and ability to care may be wrapped up in issues of social class. Though the literature on ethnicity, informal care-giving and older people is scarce in the UK, that which does exist is also briefly discussed. In addressing these issues the chapter draws on contemporary research as well as recent statistical data to examine who actually undertakes the informal care-giving role.

Gender and Care Support

The concept of care with which policy makers are concerned explicitly links care to the place in which it is performed and the social relations through which it is carried out. Informal care is seen to be a task that is performed largely (but not exclusively) within the private space of the home and through the social relationships of kin and community. Promoted as a 'labour of love', informal caring has been defined as a twofold process in which women both care *for* and care *about* (Peace 1986). Whilst women may be engaged to a greater or lesser degree in the labour market in different countries, the normative assumption underpinning the development of modern welfare states has been one in which men have been cast as having the primary responsibility for earning and women for caring. These roles were carefully nurtured through social policies that reinforced women's position in supporting a stable family life. Such policies also reflected a widespread belief that men were less able to take care of themselves – a belief Arber and Gilbert (1989) maintained, remained deeply rooted in social policies surrounding the development of care in the community. Indeed, the introduction of state benefits aimed at supporting informal carers in the UK, they argued, simply acted to buttress notions of the patriarchal family. Classified as dependents of their spouse, married women found themselves ineligible for these benefits despite the fact that they may have had to give up working or significantly reduce their paid employment in order to undertake the caring role (Peace 1986). Though this classification no longer exists, qualification for carer benefits in the UK is still means tested and dependent on: a) the cared-for being in receipt of attendance allowance, or disability living allowance; and b) being a full-time carer.

Understanding the state's role in reinforcing the gendered nature of caring is not simply a question of how social policy has been constructed, but of how it is implemented through the wider institutions of the state. Questions have been raised, for example, about the extent to which a gender bias may be embedded in care support systems in ways that act to reinforce gender inequalities in access to services. Early researchers in this field pointed to three common assumptions: firstly, that men were unlikely to be the primary carers of frail older people; secondly, that frail older men received more support from the statutory and voluntary services than equally frail older women; and thirdly, that male carers were likely to receive more support than their female counterparts (Nissel 1980). Others argued, however, that gender discrimination within the caring services was linked to women's own perceptions of their ability to care. Peace (1986), for example, maintained that this was because women were more likely to retain their social networks and be visited by friends and relatives than men (although this declined amongst very old women). As a consequence, women felt less need to ask for formal care support and care professionals were less likely to feel that these women were in need of respite services. More than a decade later, Bywaters and Harris (1998) demonstrated that gender bias in professional responses to male and female carers still persisted. In their study of UK spouse carers over the age of 75, they found that despite caring for a population with a lower dependency need overall, male carers received higher levels of support services – particularly day care, respite care and most homecare tasks with the exception of personal care. Female carers, they maintained, were less likely to be offered the support of public services than their male counterparts despite that fact that women were more likely to be undertaking a heavier caring role than men. These findings suggest a difference in expectations concerning male and female carers and the underlying assumption that women will continue caring for a longer period than men.

Echoing Peace's (1986) earlier work, Bywaters and Harris (1998) also suggested that this may be due, in part, to women's unwillingness to see their traditional role within the home being taken over by someone else. This can also be linked to women's desire to retain control of the private space of the domestic home – a view reinforced by other UK studies evident in the following interview excerpt from a female spouse carer who commented:

> I felt they [statutory services] were going to send this person in, send that person
> in – different things, and my home wouldn't have been my own! There would
> have been somebody coming in all the time, every hour of the night and day,
> and I just couldn't stand that. My home wouldn't have been my own! (Milligan
> 2000, 173).

This view of the gendered nature of choices made about the provision of personal care has been reinforced in more recent work by Arksey and Glendinning (2005). Male carers, they maintain, find intimate cross-gender caring more problematic than women; as a result men are more likely to draw boundaries around the extent of intimate care they will or will not undertake. Gender differences also extend to choices about the use of domiciliary care services, with female spouse carers more likely to reject services such as meals-on-wheels or home help, viewing them as an implicit criticism of their competence to run a home (Arksey and Glendinning 2005). Conversely, they maintain that male spouse carers are more likely to accept domestic help, possibly because it substitutes for domestic labour previously provided by their wives.

Aneshensal et al. (1995) further suggested that wives were more likely to care, and care longer than husbands, because they have a lifetime of greater experience of performing domestic tasks and caring for sick family members than men. Hence, they are better prepared to care for impaired husbands than vice versa. This view, however is at odds with current UK census data that reveals that whilst greater numbers of women undertake extra-resident or distance care than men (and thus are likely to be adult daughters) there has, in fact, been a convergence in the number of co-resident male and female *spousal* carers over the last decade or so (Maher and Green 2002). Indeed Hirst (2001 352) points out that not only did male spousal care-giving increase by around 8 percent per annum during the 1990s – nearly twice that of female spousal carers – but given spousal care is likely to increase over the next few decades as the post-war 'baby-boomers' reach their seventies and beyond, male spousal care-giving may even *exceed* that of women in future years. There is also little clear consensus about the extent to which any institutional gender bias exists. Arber and Gilbert's (1989) analysis of the 1984 General Household Survey (GHS) in the UK, for example, indicated that after controlling for the level of frailty, the amount of support services received did not in fact vary by gender, but rather by household type. That is, as would be anticipated, lone elderly households were likely to receive greater levels of formal support than those with a co-resident spouse or other family member, irrespective of gender.

With the exception of the work by Arksey and Glendinning (2005) limited attention has been paid to the extent to which gender inequalities in access to formal care services still exists in the UK, hence there is little data against which such findings can be compared. Though the UK census and the General Household Survey both gather data about informal caring, gender and services received, the responses to these questions have not been correlated, hence there is no clear picture of whether or not gender differences in the type or extent of formal services carers receive exist. Further, whilst we know that there are regional variations in the *extent* of informal care-giving across the UK (Maher and Green 2002), it is unclear whether these variations are gendered and how this may or may not relate to regional variations in access to formal services. So while the findings that emerged from the Bywaters and Harris (1998) study are interesting, and point to an urgent need for further investigation into these issues, they need to be read with caution. As they, themselves openly acknowledged, their findings were based on a fairly small-scale project located within only one social services department in the UK. Without more detailed analysis, it is difficult to tell whether such biases are locally specific or whether they may be endemic throughout the system.

Recent research in the Canadian context however, suggests that there may, indeed, be a gendered pattern to the forms of caring activity undertaken by men and women (e.g. Campbell and Martin-Matthews 2000 2004; Williams 2006). Campbell and Martin-Matthews maintain that men are more likely to involve themselves in care that is tied to traditional male roles in the family, such as home maintenance, financial and administrative tasks than the more intimate care tasks such as bathing, toileting, dressing and so forth that are often undertaken by women. Interestingly, Matthews' (2002) work on the gendered nature of care work performed by adult sons and daughters in the United States suggests that the more proactive approach taken by many women as they seek to pre-empt parental needs and requirements, may in fact act to *increase* dependence as such actions reduce the need for ageing parents to make the attempt to undertake these tasks themselves. Conversely, they maintain that adult sons are more likely to take a more reactive approach to the performance of care that in turn is likely to promote their older parents' independence. These are important considerations – ones that warrant further attention.

Where adult children have siblings, researchers have pointed to gendered and structured inequalities in their ability to exert agency in their relationships with one another and other family members (see Connidis and Kemp 2008). These inequalities are often based on intra-familial gender and class relationships. In particular, though both adult sons and daughters may provide care to an ageing parent, as suggested above, women often provide the bulk of the direct care. Sons, on the other hand, are often the higher income earners and as a consequence can find themselves being pressured to provide more economic support. Alternatively they may use their additional economic resources to buy in care support in lieu of providing personal support themselves. For Connidis and Kemp (2008) this can be interpreted as part of what they refer to as the 'legitimated excuses' that exist within the framework of 'distributive justice' (231) that underpins familial networks of care. Here, distributive justice is linked to the way in which familial contributions to care are apportioned based on the successful claiming of legitimated excuses. Within families, issues such as proximity and distance, gender, the 'taboos' of cross-sex caring, employment, relationship history, other family dependencies, marital and health status can all influence which siblings are expected to provide care and who is 'legitimately' excused. Contributions to care are thus viewed as equitable based on the proportion of a family member's ability to contribute.

In deciding who should undertake the primary caring roles within family networks, Campbell and Martin-Matthews (2004) also point to a 'hierarchical pattern' of informal care-giving that is headed by spouses then adult daughters followed by adult sons and other family members. Adult sons who are primary care-givers, they maintain, are less likely to have siblings and as a result find themselves 'caring by default' (Campbell and Martin-Matthews 2000, 109). Where men are the primary caregiver, particularly where they are co-resident carers, traditional gendered caring roles can break down as men are no longer able to claim a 'legitimating excuse' (Campbell and Martin-Matthews 2004, 352).

Hence, understanding the extent to which gender discrimination in care support exists cannot solely be confined to the domain of the formal care service. Early commentators suggested that not only are male carers more likely to gain additional informal care support from other friends or relatives than female carers, but they are also more likely to be provided with it at an earlier stage in their caring career (Charlesworth et al. 1984). Maher and Green's (2002, 12) analysis of the 2000 census data in the UK goes some way toward supporting this claim in that they note that women are more likely to be the sole main carer than men (23 percent to 16 percent) and significantly more men are likely to be secondary carers than women (43 percent to 32 percent). Based on such analyses Chamberlayne et al. (1997) were lead to conclude that women are caught up in a web of complex familial relationships that incorporate hidden gendered pressures linked to conflicting ties of obligation and affection. In other words, adult daughters and other female relations were more likely to be (or feel) pressured into caring than their male counterparts.

This, then, points to the need for a closer examination of the caring relationship, as intrinsic gendered attitudes towards the provision of both formal and informal care support may act to reinforce the notion of caring as a gendered concept – one that positions women firmly at the centre of the caring role.

In the UK, the 1985 GHS Survey on Informal Carers represented a landmark as the first attempt to collect information on what had previously been 'hidden' domestic labour in the UK. Subsequent surveys undertaken in 1990, 1995 and 2000, together with the inclusion of questions around informal care in the 2001 Census, are beginning to give us a better picture of who cares, where and the extent of informal care they undertake. Yet the gathering of such data is not unproblematic. Firstly, it requires individuals to self-define as carers; and secondly, it assumes that caring involves undertaking extra domestic responsibilities. While this may be relatively unproblematic for non-resident carers where extra responsibilities can be easily defined, this is less clear-cut in the case of co-resident spousal care. For example, while older men are likely to include undertaking cooking or shopping tasks for a dependent wife as 'extra tasks' – an elderly wife is unlikely to define them as additional responsibilities in caring for a dependent husband (Arber and Ginn 1990). Equally, older women may take on additional tasks such as basic house maintenance (mowing the lawn, painting, minor house repairs etc.) when caring for a dependent husband – tasks an elderly male is less likely to define as additional responsibilities.

Gender and care within residential care settings

Policies designed around ageing in place promote the ideal of family care within the domestic home. Home, with its connotations of private and family life brings us full circle to the domain of caring women. Yet the domestic home is not the sole locus of caring for older people. A significant minority of frail older people are likely to spend at least some of their final years in residential care, and here too the gendered pattern of care is markedly different.

An early study by Willcocks et al. (1982) illustrated that not only did women outnumber men by 3:1 in residential care homes, but that only 26 percent of these men were over the age of 85 compared to 44 percent of women (73). Hence men were seen to enter care homes at an earlier age and were generally in better health and more mobile than women. More recent ONS data (2001) indicates that there has been little change in this pattern, with women making up 76 percent of those aged 65 and above who are in residential care. Willcocks noted that bereavement was often a common factor in male entry to care homes, suggesting that older men may be less able to care for themselves than women. Peace (1986) further suggested that care home entry may be less traumatic for older men than older women as (at least for current cohorts) they may already have experience of communal living (i.e. through national service etc.) and the predominance of women in the formal care-giving sector means that female carers at home are largely replaced by female carers in the care home. For women, used to the private space of the home and their status as the main carer within the home, this shift to the 'semi-public' space of the care home may be more traumatic. Their traditional domestic skills are neither needed nor recognised, impacting on their self-esteem. While the older woman continues to live in her own home, she still exercises a degree of power and control over her life, but once located within the institutional environment, this power base crumbles away to be replaced by dependency and subordination. Indeed, Peace suggests that this experience may be further aggravated where the older woman has been supported and cared for within the domestic home by an adult daughter and feels let down, resulting in feelings of anger or frustration.

Changing Patterns of Work, Family and the Gendered Nature of the Caring Relationship

Seventy percent (nearly five million) of all informal carers in the UK care for someone over the age of 65 (Maher and Green 2002). Though women are still more likely to be carers than men – with 56 percent of all carers in Britain being female compared to 44 percent males – as noted above, the extent of male care-giving is surprisingly high (Carmichael and Charles 2003; Dahlberg et al. 2007). Though the gender divide in spousal care-giving has gradually disappeared, there has nevertheless been an increase in the extent and intensity of extra-resident care delivered by daughters within the parental home. Women in the UK, for example, are 50 percent more likely to care for parents or parents-in-law than men (Hirst 2001, 353). But whilst women have increase their involvement in familial care, the data reveal that they have also reduced their involvement in less intensive caring relationships (e.g. amongst friends and neighbours), pointing to a gendered intensification of care-giving (Hirst 2001). In part this is linked to women's increasing participation in the labour market and their need to prioritise how, and to whom, they allocate their time outside of the work environment; and in part to shifts in home ownership patterns and older people's preference to retain as much independence as possible.

Across Europe, around three times as many women devote a substantial part of their time to the performance of informal care than do men, though substantial differences exist across nation states (Bettio and Plantenga 2004). These differences can be linked to a number of factors. In countries with liberal or conservative welfare regimes the family responsibility to care is still largely devolved to women. Indeed, data from the European Community Household Panel on the volume, character, and intra-household distribution of informal care, together with data for pre-accession countries points to a strong link between care systems and the female labour market. Bettio and Plantenga (2004, 104) maintain that in countries such as Italy, Spain and Greece, where care tends to be a family responsibility, 'options for reconciling work and women's care tasks are relatively scarce, costly, or perceived as offering inferior-quality care, with the result that many women take responsibility for housework and/or care work instead of seeking paid work'.

Current demographic trends in many advanced capitalist countries also point to the emergence of a 'sandwich generation' of parents raising and supporting adult children who are likely to remain dependent for longer whilst also having older parents in need of support (Grundy and Henretta 2006). This 'sandwich generation' is viewed as arising as a result of women childbearing in later years, younger generations spending more years in education and training – so remaining dependent for longer upon their parents – and the ageing of the population. Jönsson (2003) maintains that these new multigenerational kinship patterns are likely to result in more adults experiencing a 'middle-generation squeeze' when they may find themselves caring not only for grandchildren but also for their frail older parents.

Increasing divorce rates and cohabitation are also creating a rise in the number of 'step-families'. These shifts are changing the quality of intergenerational relationships raising questions about their ability and willingness to care for frail and older step-parents who have no direct blood kinship tie. Though adult children in a step-familial relationship still acknowledge the potential need to support their older step-relatives, they do so in a way that prioritises their own immediate familial needs first (Bornat et al. 1999). Campbell and Martin-Matthews (2000) further maintain that while blood ties still underpin the availability and willingness to care within these new extended families, how relationships are renegotiated at the time of divorce or separation are important in determining the nature and quality of care and support given. Though the significance of gender within the caring relationships that emerge from these new extended families has yet to be determined, it seems unlikely that there will be any significant shift from the patterns described above.

Care-giving as a Gendered Politics of Place

To fully understand the gendered dimension of the care-giving experience, it is also necessary to appreciate the structure of the welfare state within its national context. As discussed in Chapter 2, different welfare regimes have different implications for how care is constructed and who is seen to have the primary responsibility for care. While the 'male breadwinner' model, with its implications for gender and family structure, may have underpinned the development of the modern welfare state, this approach to care is changing (Lewis 2002). Commentators point to two key drivers: i) external forces, such as the impact of globalisation; and ii) internal forces such as demographic and economic change (e.g. Esping-Anderson 1999; Daly and Lewis 2000). The changing nature of the family is integral to the latter point and is not just associated with shifts in the family structure, but also low birth rates and the ageing of the population. Gender is seen as instrumental to the solutions proposed to accommodate these changes. More specifically, it has been suggested that care work should be shifted to the public/market sectors in order to facilitate women's employment (Esping-Anderson 1999). Lewis (2002) argued, however, that to do so without considering issues of gender and contestations around care is highly problematic.

There has also been a sustained effort in modernising welfare states to recast social welfare within a framework of rights and responsibilities – with an emphasis on 'active' rather than 'passive' welfare, albeit to a greater or lesser degree within different countries. Such an approach places an emphasis on the 'worker citizen' model (Lewis 2002), emphasising paid work for all adults (male and female) with a concomitant commodification of labour (Esping-Anderson et al. 2001). This runs counter to the traditional 'male breadwinner' model of earlier welfare regimes – a model that some neo-liberalising countries recognise 'is increasingly out of date' (DSS Cmd. 3805 1998, 13). Yet, as Lewis (2002) has cogently argued, the promotion of the worker citizen model fails to engage with the implications for the gendered division of labour – particularly the unpaid labour of informal care. Bettio and Plantenga (2004) go so far as to claim that despite attempts to invigorate a citizen worker model, existing barriers to women's employment, reinforced by their high levels of participation in informal care, are also likely to increase the risk of gendered poverty in old age. That is, as women tend to outlive men, they are more likely to find themselves widowed with minimal resources (Fagan and Burchell 2002). While there have been some instrumental changes to compensate for the 'worker citizen' model in the UK – such as new pensions for carers, unpaid leave for family emergencies linked to care and increased respite provision – where such changes have been implemented they have been relatively weak (Carmichael and Charles 2003). Hence as Lewis comments, 'the fate of women in respect of the new work/welfare relationship depends in large measure on what provisions are made for what was the unarticulated dimension of the traditional male breadwinner model: unpaid care work' (2002, 345).

While the predominant approach across modernising welfare states has been one of seeking to reconcile work and family life, this inherently entails a feminisation of the workforce and a concomitant commodification of care – and paid care is still disproportionately undertaken by women (Carmichael and Charles 2003). The gendered nature of the generally low-paid care labour market is thus likely to compound gender inequalities in care. Yet care entails not only those active care-giving tasks such as domestic and personal care and medication, but also passive activities, such as simply being there – any attempt to commodify all such care work would thus be extremely difficult.

Such arguments are framed around informal care-giving in its broadest sense. That is, it incorporates those forms of informal care-giving that include childcare, caring for younger mentally impaired or physically disabled people, as well as informal care work with frail older people. In considering the impact of change in relation to the care of frail older people, such claims are clearly significant in terms of intergenerational care-giving. Yet we should not forget that the largest proportion of those providing informal care to frail older people are spousal care-givers. For them, such debates will be of limited significance. It is only when spousal care-giving breaks down, or there is no spousal care-giver, that the impact of any shift to a 'citizen worker' model of welfare will be of significance.

Class and Caring

> All women are expected to provide informal care, but how informal caring affects them is likely to vary by class (Arber and Ginn 1992, 927).

In attempting too tease out the thorny issue of who cares, where and why, it is important to consider the extent to which class may influence the informal caring role. While there has been a relative abundance of work that has conceptualised informal care-giving within a gendered framework, far less attention has focused on the influence of class. Perhaps the most significant work in this field in the UK is that undertaken by Jay Ginn and Sarah Arber (1992) despite the fact that this work is now some seventeen years old. Drawing on GHS data they were able to demonstrate that working class men and women exhibit significantly different perspectives toward informal caring than do middle class men and women. Combined with material disadvantage, these differences may result in different patterns of care across the class divide.

Whether a frail or disabled older person requires informal care is not simply a function of their level of physical impairment, it is also mediated: firstly, by those material, financial and cultural resources on which they can draw; and secondly, by the extent of available state provision and resources. Those with the financial resources to do so can use these resources to purchase aids and adaptations over and above those provided by the state. This can facilitate their ability to live independently within their own home as well as pay for additional privately purchased care support. Caldock (1992), for example, found the purchase of private help by frail older people to be exclusively the preserve of the middle classes. Furthermore, while impairments in later life are often considered to be an almost inevitable part of the ageing process, class inequalities still exist. The more affluent middle classes, for example, experience lower levels of disability than those in the working classes – a prevalence that persists even amongst those in their eighties (Arber and Ginn 1992).

While these analyses seem to point squarely toward a class-based difference in the care experience of older people, it is important that any analysis of this kind also take into account class issues in relation to the informal care-giver – particularly non-spousal carers. The financial and material resources of the non-spousal carer can be important in influencing not only who cares, but the quality of care given. Those with the financial resources to do so may also purchase additional aids, adaptations and services that enable the care-recipient to maintain their autonomy for longer. Material resources, such as car ownership, can also facilitate the provision of extra-resident care, reducing the pressure for co-residence and extending the independence of the cared-recipient. Informal carers also play a crucial role as intermediaries working on behalf of the care-recipient to negotiate a preferential set of services – a role that can be performed whether the informal carer is proximate or distant. Class differences are also important here in that knowledge and educational status can affect the ability of the informal carer to negotiate with formal service providers. Informal carers with sufficient financial resources may further choose to purchase residential care on behalf of the cared-recipient, and in doing so, reduce the potential pressures and constraints on themselves as care-givers.

Yet even with the added class element, commentators such as Arber and Ginn (1992) and Campbell and Martin-Matthews (2004) note that that there is a gendered dimension to the care-giving experience. Not only are working class women less likely to be able to present 'legitimating excuses' related to the opportunity costs of employment relative to those of working class men, they are also disproportionately disadvantaged by the burdens of informal care in comparison to women from middle class backgrounds. In part, this is because older working class men are more likely to experience poorer health (and at a younger age) than their middle class equivalents. Indeed, while working class men are likely to require care in their sixties, middle class men are more likely to be in their eighties before requiring informal care and support (Maher and Green 2002). The implications of this age difference in the likely need for informal care has two significant consequence for non-spousal carers: firstly, the material advantages enjoyed by older middle class people (or their carers) means they are more likely to be able to live independently in their own homes for longer; and secondly, non-spousal carers from the most materially disadvantaged families are more likely to be caring for other sick or disabled family members, reducing their ability to care at a distance. This group of carers then, is more likely to undertake co-resident caring, and co-resident carers are more likely to experience adverse impacts to their own health as a result of their caring role (Maher and Green 2002).

One further observation is that given, as previously indicated, that non-spousal carers are more likely to be women, where intensive caring resulting in an exit from the labour market is undertaken at a younger age, the implications for the re-establishment of their careers following disengagement from the caring role is likely to be more profound. Carmichael and Charles (2003) point out that while paid employment for both male and female carers declines as the caring role intensifies, not only are women more likely to be the main carer and hence spend more hours caring than men, but male carers are also more likely to maintain a stronger attachment to the labour force. On the whole, they maintain, men do not willingly give up paid work even when caring for someone with high levels of support needs. Inevitably this has a knock-on effect in later life as older women who have undertaken informal caring at a younger age are less likely to have the financial benefit of a work-related pension than their male counterparts.

Research also points to informal care becoming increasingly economically differentiated *between* women. Jönsson's (2003) work on informal care in Europe revealed, for example, that in southern European countries, as with their male counterparts, well-educated women with high incomes and secure employment often bought-in services aids and adaptations for older relatives rather than choosing to undertake direct caring themselves. Moreover this paid care work was often undertaken by less educated, low-paid women either from their home country or from migrant labour drawn from other parts of the world. Jönsson is thus lead to suggest that more educated and affluent women may, in fact, be contributing to and reinforcing gender inequalities in care. Conversely, families with lower incomes continued to rely on female relatives, friends or neighbours. It is worth noting however, that whilst undoubtedly the purchase of care services by women may exacerbate gender and socio-economic inequalities in care, in the UK at least, the extent to which this can be attributed solely to more affluent women is likely to become increasingly muddied. The introduction of Direct Payments

Schemes in 2000, for example, has enabled all older people assessed as being entitled to care support to receive a cash sum to purchase the services they prefer rather than relying on state provision. Given that care support is means-tested, this means that those taking up this option will in fact be drawn from lower socio-economic groups. Whilst the numbers of older people taking up this option are as yet relatively low, they are increasing (NHS Health and Social Care Information Centre 2006). Hence in the UK at least, women carers from both ends of the socio-economic spectrum are likely to contribute to an exacerbation of gender inequalities in paid care work.

In sum, while informal carers may be relatively equally drawn from all classes, where, when and how informal care support is given can differ significantly. Middle class carers have more options and more leverage in negotiating with social welfare professionals to obtain public services or residential care than those from working class backgrounds. They are also more likely to provide extra-resident care, using either their own, or the material resources of the cared-for to support the older person's ability to remain independently in their own home for longer. Co-residence is more prevalent amongst the working classes, and given that co-resident care tends to be more intensive, it is likely to impact more on the life of the informal carer, and at a younger age. Carers from different class backgrounds, thus, appear to manage the needs of the cared-for in significantly different ways. The extent to which older people require informal care is thus, to some extent, inextricably bound up with class.

Ethnicity, Culture and Care

Feminist approaches to the analysis of care have been useful in facilitating our understanding of how care has been socially constructed as a gendered concept that reflects a specifically 'feminine' expression of society. However, as Graham (1991) cautions, such analyses have, by and large, tended to be been both culturally and ethnically 'blind'. In its efforts to identify carers as a gendered social group sharing common problems and interests, with a few notable exceptions, feminist research has tended to be one-dimensional in regard to different forms of care and the social divisions of women's experiences of caring at home. Such an approach fails to recognise ethnic and cultural differences in care and responses to care. Hence, their experiences are likely to be shaped by an absence, rather than a presence, of a clearly defined private sphere. Yet as Figure 3.1 illustrates, in the UK a significant proportion of individuals from Black and other ethnic and cultural groups provide care. Indeed the proportion of Indian people in Britain providing unpaid care is not dissimilar to that of the white British population.

The pattern of informal care-giving across different cultural and ethnic groups in the UK varies, with those from mixed ethnic backgrounds (5.1 percent), Black Africans (5.6 percent) and the Chinese (5.8 percent) being least likely to perform informal care. To some extent, this reflects different age structures within and

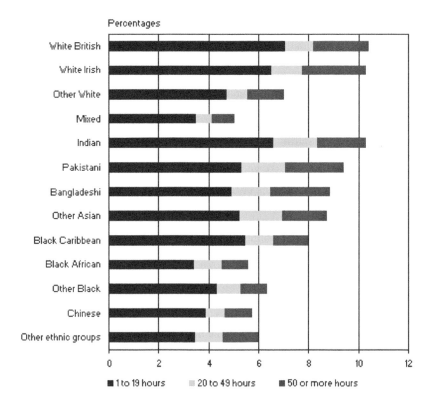

Figure 3.1   Care by ethnic group and time spent caring in the UK (2001)

Source: Office for National Statistics 2001. Reproduced with permission.

across different ethnic groups, as informal care is most likely to be provided by people aged between 50-60 years of age (Maher and Green 2002). But it is also shaped by different cultural factors. Graham (1991) for example, pointed out how Black women's work outside the home has historically taken precedence over the needs of their own families.

The amount of time that people spend caring in the UK also differs by ethnic group. Those most likely to provide substantial amounts of time caring for their older relatives tend come from the same groups who provide care in the first place. Bangladeshi (2.4 percent) and Pakistani (2.4 percent) groups are more likely to spend the most time caring (50 hours a week or more), with Indian, Pakistani, Bangladeshi and other Asian groups most likely to spend 20-49 hours a week caring (1.5 percent or slightly more for each group) (ONS 2001).

Despite clear evidence that a significant amount of informal care is undertaken by individuals from ethnic and cultural minority groups in the UK, there is very little research on their experiences of caring (Koffman and Higginson 2003;

Merrell et al. 2006). Yet as those researchers who *have* undertaken work in this field point out, for many older people from different cultures, the rituals and traditions associated with their culture can become more important with advanced age and increasing frailty. Merrill et al.'s (2006) work with Bangladeshi carers in Wales, for example, demonstrated that not only do formal services tend to be ethnocentric – attempting to imbue ethnic communities with western values rather than trying to understand them – but formal service providers tend to assume the stereotypical view that large extended families in these communities will willingly and unquestionably contribute to the provision of informal care to their ageing populations. But as Blakemore (2000) notes, in reality, as with the white British population, changing family structures and residential patterns are breaking down these traditional familial patterns of care.

Shifts in traditional family patterns of care are not the only factors affecting access to care amongst ethnic minority groups in the UK. Murray and Brown's (1998) work with Black and ethnic older people across a range of Local Authority areas with high minority ethnic populations in England also highlighted the confusion and difficulties carers from these groups can face in trying to understand the highly complicated carer benefits system. Merrill et al. (2006) also point to inequalities in access arising from the difficulties many first generation migrants have with written and spoken English. Understanding the complex pathways to care in the UK is a challenge for even the most informed of informal carers, but this becomes even more so when compounded by a lack of fluent spoken and written English. Language problems can be both age and gender related (Nazroo 1997; Gerrish 2001; Merrell et al. 2006). South Asian men, for example, are more likely to able to communicate in English than their female counterparts, though younger women (particularly those in their teens and mid-twenties) are more likely to speak English. The language barrier is thus a particular issue for older female carers, often rendering them reliant on children, other friends or relatives to help them negotiate the complex care pathways. As a result, they can often miss out on benefits or services (Merrell et al. 2006). Indeed, the take up of financial support due to carers is often substantially lower in areas with a high proportion of minority ethnic populations than other areas (Rosato and O'Reilly 2006). In other contexts such as North America and Europe, commentators have noted that failure to access welfare benefits amongst minority ethnic groups in is often based on traditional (largely gendered) expectations of responsibilities for caring for frail older relatives. This can be reinforced by experiences in their home country where not only may welfare support for informal carers not have existed, but the very term 'carer' may not exist in the language of origin (Lan 2000; Warnes et al. 2004; Team et al. 2007). As Team et al. (2007) point out, it can take several generations for these dominant attitudes to change.

Though work on ethnicity and caring for older people in the UK is limited, work undertaken in other advanced capitalist countries has drawn attention to some important differences in social norms. These are linked to care-giving and attitudes to care across different racial and cultural groups that arise due to differing experiences, norms, and attitude to healthcare (see Bradley et al. 2004). Team et al.'s (2007) work with Russian migrant women carers in the United States and Spitzer et al.'s (2003) work with Asian and Chinese migrant women carers in Canada, for example, point to the importance of the care-giving role to these women in both reaffirming their gendered and (in the case of the Russian women) their religious identity. Spitzer et al. also highlighted the central role that this care work can play in maintaining and reproducing migrant women's 'ethno-cultural community'. Such work makes an important contribution to our understanding of the relationship between ethnicity and informal care-giving within specific ethnic migrant communities. Moreover, in the Canadian context, Spitzer (2003) noted that South Asian women are not only typically employed, but have little room to negotiate their care work responsibilities within families from whom they receive little support. Neoliberal policies and the retrenchment of state support for health care is not only exacerbating the demands of the care work that migrant women perform, but employing paid care workers to ease the burden is made virtually impossible, not only because of the low incomes they earn, but also because their elderly relatives resent workers from outside the family circle (Bradley et al. 2004). At the level of individuals and the home, cultural norms about the role of family in providing care and the need to maintain autonomy, can lead to resistance to seeking – or allowing women to seek – formal care particularly where this may mean care workers enter the home (Wiles 2005). The confluence of these factors makes gendered care work in Canada more costly for these women than it would have been in their countries of origin. Hence, unpacking the power relations connected to culture and ethnicity is critical to any real understanding of how care and healthcare for older people is experienced.

Concluding Comments

Whilst informal care forms the lynchpin around which formal community health and social care policies for frail and disabled older people are formed, the feminisation of the workforce, increased geographical mobility and changing family structures are impacting on the availability of this informal care. Whilst there has been an increase in male care-giving in the UK, as yet this is linked only to spousal care-giving within the home. Indeed not only are women more likely than men to care for someone outside the household, but they also predominate amongst those sub-groups that undertake the heaviest caring commitments (Maher and Green 2002). Gendered inequalities in care are exacerbated for women from lower socio-economic groups and from ethnic minority backgrounds both as a consequence of poorer health and of heightened difficulties in accessing care support. Furthermore, amongst women from both lower socio-economic and ethnic minority groups, it is clear that differing cultural norms act to reinforce the gendering of care.

Chapter 4
Mapping the Contours of Care –
International and Transnational Perspectives

Definitions of what constitutes care and the extent of informal care-giving are highly cultural (Tronto 1993; McDaid and Sassi 2001). Yet as the discussion around welfare regimes in western society in Chapter 2 has demonstrated, who cares and where that care takes place is also subject to varying perceptions of rights and responsibilities. Care, then, is shaped not only by cultural practice but also by political and economic circumstances that vary across space. This aim of this chapter is to consider how some of these dimensions of care are manifest within non-western societies. Its global coverage is by no means comprehensive, indeed, it would be impossible to do so in the space of one short chapter, rather the aim is to tease out some of the key factors that contribute to cultural and cross-national differences in the construction of care for older people and how these are affected by contemporary developments in society that operate across a range of scales from the local to the global.

When considering cross-national differences, it is also worth noting that what constitutes an 'older person' is not clear-cut. While 60 years of age is typically taken as the line dividing older and younger age cohorts in demographic analysis (UN 2007), many affluent societies tend to classify old age in terms of pensionable age. This is defined by retirement legislation that is regulated by chronological age and chronic disease. Typically, this has referred to people aged 60-65 years, but this is gradually shifting as increased life expectancy, combined with increasing pressures on pension and welfare systems in some developed countries, has manifest in a move towards extending the pensionable age (Turner 2005). Apt (2002, 40) notes, however, that 'In the poorer populations of the world [old] age is associated largely with physical limitations, the inability to work and economic dependency'. Though we are seeing a global ageing of the population, older people in most developing countries still have shorter life expectancy than those in developed countries. Retirement legislation or state support for older people in these countries is either negligible or non-existent. Older people, then, must either rely on alternative forms of support, or continue to work until they are no longer physically able to do so or die (Ssengonzi 2007). These observations suggest the need for a more fluid concept of the 'older person' – one that is defined less by chronological age and state legislation and more by frailty and reliance on support to undertake the activities of daily life. The extent of this reliance may change over time and vary across social and spatial contexts.

Globalisation and the Moral Economy of Care

Geographers have engaged with issues of care, and the responsibility to care, for over two decades (Lawson 2007). This engagement, however, has largely been concerned with universal notions of care – drawing, for example on early debates within moral philosophy about the gendered nature of care ethics (see for example, Kohlberg 1981; and Gilligan 1982). These debates focused around universal issues of good and right that were seen to reflect a male view of the world (viewed as rational and dispassionate) versus the home. As an expression of female space, the home was characterised as a private place of nurture and emotion, and as such, beyond the reach and protection of social and legal justice (Smith 2000). Friedman (1993) argued, however, that any concern with issues of care, nurture and the maintenance of interpersonal relationships shifts care ethics *beyond* the private realm, giving it the status of basic (universal) moral importance.

The spatial scope of care and beneficence has been of particular interest to geographers. A spate of work has emerged from the 1990s onwards that considers care through an engagement with ethical and moral dilemmas as well as issues of responsibility and social justice (see for example, Harvey 1996; Smith 1998 2000; Silk 1998 2000; Proctor and Smith 1999 2004; Massey 2004; Smith 2005). Within welfare states, welfare retrenchment, coupled with the market logics of competition and efficiency are perceived to underpin the imperative for cutbacks in care support – and in ways that impact disproportionately on the poorest in our societies. For those concerned with the wider responsibility to care, these shifts are manifest in a geography of poverty and inequality that reflects who has access to care and who undertakes that care work. Within this frame of reference, the spatiality of care is interpreted as an ethical concern emerging from the so-called 'moral crisis' that threatens contemporary western society. Society, Smith maintained, has 'lost its moral bearings' (2000, vii) as it fails to get to grips with the increasing polarisation between rich and poor; growing intolerance to difference (whether socio-economic, religious or culturally-based); and the pursuit of personal improvement over that of wider society.

Debates about the moral imperatives to care within a global context are clearly important in that they address the growing divide between rich and poor at an international level and exhort us to accept a responsibility to both care for and about distant others. These debates are located around universal notions of care that draw on issues of difference, beneficence, inequality, citizenship, rights and responsibilities (Milligan et al. 2007). They consider the ethics of care across a range of spatial scales stretching from the global to the highly place-specific. While they play a critically important part in helping us to understand and address issues of ethics, morality and social justice we should not overlook the fact that, while distance does not preclude the performance of material care (as outlined in Chapter 2), at its most fundamental and intensive level, caring *for* involves the personal and the proximate. This in turn is shaped by the differing political and socio-cultural circumstances within which care takes place.

Care, Culture and Change

The most striking feature of traditional care systems in Africa, Asia and Latin America is that they are rooted in complex family systems that include reciprocal care and assistance amongst the generations (Van Dullemen 2006). Care is thus defined as an almost exclusively private activity built around notions of familial obligation and is both given and received within the domestic space of the home. With a few key exceptions, the state either fails to provide any form of social support or provides only residual support aimed at the most destitute (Chan 2005; Glaser et al. 2006). Within these family-based systems, older people requiring care and support receive it within an extended family network, but at the same time they, themselves, can also undertake an active and reciprocal care-giving role. Child fostering by grandparents and other older relatives in Sub-Saharan Africa for example, (albeit in different forms across the continent) is a common phenomenon designed to share the costs of childrearing and strengthen kinship ties (Ssengonzi 2007). Traditionally then, the family has been the greatest site of security for its oldest members, but at the same time, older people form an active and integral part of the fabric of civil society.

Despite the impact of the very differing socio-cultural and political landscapes of care, the World Health Organisation (WHO) (2002) points to two key changes that, globally, reflect a steep increase in the need for long-term care:

i.   The growing prevalence of long-term disability in the population; and
ii. Changes in the capacity of informal care support systems to address these needs.

It is often assumed that these are phenomena that pertain largely to developed countries. But long term care needs in both transitional and developing countries are increasing at a rate that far exceeds that experienced by developed countries. Indeed, a recent United Nations report (2007) estimates that if current trends persist, by 2050 almost 80 percent of the world's population of those over 60 years of age will be living in what are now developing countries. The WHO further point out that not only is the developing world ageing at much lower income levels than those that characterised the same demographic transition in the developed world, but it also has much lower levels of income with which to meet these emergent needs.

Whilst local cultural and structural factors play an important role in how care for older people is manifest within both developing and transitional countries, a number of common issues are also affecting who cares and where that care takes place including:

• The effects of modernisation and urbanisation;
• Shifts in traditional family structures;
• Changing cultural and ethical values;

- The evolution of health and social welfare systems; but also
- Limited available resources with which to develop effective welfare programmes.

Shifting population patterns, particularly global ageing, means that these countries are likely to face future long-term care needs that will exceed the experience of the developed world.

Sub-Saharan Africa

Even though most older people requiring care in developing countries are still looked after within the informal structures of the family, as we move forward into the twenty-first century, these family-based systems of care can no longer be taken for granted (Apt 2002). Population ageing, combined with modernisation, is having a significant impact on who cares where. Projections indicate, for example, that by the middle of the twenty-first century there will over 102 million older people in Sub-Saharan Africa – 22 million of whom will be over 80 years of age (Van Dulleman 2006, 101). At the same time, modernisation and rapid urbanisation are changing traditional family structures, and hence the availability of informal care (Apt 2002). The migration of younger adults to urban areas in search of work and better lifestyles is acting to break-up the traditional extended family. As a consequence, rural areas are becoming increasingly populated by a residual population of older people with no visible means of care support.

The disintegration of traditional family structures is not solely due to the impact of modernisation. In the African sub-continent the impact of the AIDS epidemic has been particularly insidious. Despite epidemiological transition from the 1970s onwards, work by Robson (2000) and Ssengonzi (2007) in Zimbabwe and Uganda respectively, has demonstrated how high levels of AIDS infection amongst young mothers and working age men has resulted in an increasing burden of the care-giving role being placed on older women and young girls. Older women in particular can find themselves undertaking multiple caring roles, caring not only for a frail ageing spouse or parent, but also for siblings and adult children with AIDS and other chronic illnesses. This is exemplified through the words of one female carer in a study by Shaibu and Wallhagen (2002 144) in Botswana who commented, 'I have five children of my own, two grandchildren, four siblings, my mother, my elderly aunt, and I am solely responsible for all of them'. In the current context of the AIDS epidemic, and with many younger women leaving rural villages to seek work in urban areas, familial care-giving in Sub-Saharan Africa is at a premium.

Livingston maintains that, 'care-giving provides a public index of the strength of positive familial relationships and is central to the moral economy of ageing' (2002, 222) – but that that moral economy is in flux. The combined influence of global ageing, the AIDS pandemic and modernisation in developing countries means that care-giving by the extended family is being placed in an increasingly precarious position. In some countries this is exacerbated by the continuance of armed conflict, the rise of road accidents and injuries and the continued impact of tropical and other communicable diseases (Apt 2002). The precariousness of extended family care networks is such that in some areas of Africa older people, particularly women without traditional family care networks, have begun to organise themselves to provide alternative networks of care. Apt (2004) for example, cites instances of older women (all over the age of 65) each of whom also care for grandchildren, working together to find income generating projects in order to help each other live in dignity – especially very frail or bedridden older people. Hence, in some cases, older people in developing countries are now caring not just for their own children or grandchildren, but for other older people too.

The dual impact of modernisation and epidemiological transition combined with declining infectious disease and rising chronic debilitating illness in older age is not only creating changes in who cares, it is also creating socio-cultural transformations manifest in a dismantling the traditional gerontocratic hierarchy in some societies. The ability of older people to command respect, power and familial care is breaking down. Livingston's (2002) work with Tswana society, for example, illustrates how previous cultural norms that required daughters to care for mothers in old age and mothers to provide care for their sick/disabled daughters have broken down. Indeed, the growth of chronic and disabling health conditions at an earlier age has resulted in older Tswana women seeking to redefine chronic illness in their adult children as a physiological condition attached to old age. By redefining health conditions in this way, the 'burden of care' for the adult child is taken out of the hands of the older mother and placed firmly in the hands of grandchildren. Old age in this society is thus being consciously reconfigured to shift the burden of responsibility to care.

Changes of this nature place pressures on the state to develop some system of social protection for older people. Yet whilst modernisation may be increasing pressure for the expansion of state care in developing countries, this is being counterbalanced by reduced government spending on care provision as the these countries struggle to deal with the impacts of 'Economic Structural Adjustment Policies'. As a consequence, those services that *do* exist are often irregular, limited, or inaccessible (WHO 2000).

With few formal care programmes and institutional care seen as culturally inappropriate, female children and women often perform far greater levels of care than would be expected of informal carers in developed and middle-income countries. Young girls are commonly sent to live with and support ageing grandparents. They are also the first to be taken out of school to support the primary care-giver resulting not only in social isolation, but also the compromising of their opportunities for future employment. These informal carers also find themselves dealing with care issues such as incontinence, diarrhoea, bedsores, severe pain etc. without the labour saving devices that many in the developed world take for granted (for example indoor sanitation, washing machines etc.) and without the relative ease of access to services (e.g. through public or private motorised transport).

Whilst intimate care is seen as the exclusive responsibility of the family, some aspects of care – particularly intimate care – can also be culturally proscribed. Tswana society, for example, has gender proscriptions that render it difficult or taboo for women to bathe their fathers and for men to launder their mothers' underclothing (Shaibu and Wallhagen 2002). Compounded by high levels of poverty, many elderly women carers become overwhelmed by the magnitude and multiplicity of the caring tasks they are required to perform. Malnutrition, depression and personal neglect is a common outcome. In Botswana alone, 30 percent of families living in urban areas and 64 percent of families in rural areas live below the poverty line (Shaibu and Wallhagen 2002, 140). Yet despite this, they note that the stigma attached to being classified as qualifying for the Destitute Programme aimed at relieving the very poorest elderly, means many prefer to struggle on rather than apply for relief.

National conditions, values and culture, together with existing heath and care policies and practices and the impacts of modernisation and epidemiological transition mean that how informal care and the capacity to care is manifest, and what this means in terms of future care needs varies significantly across Sub-Saharan states. Yet while traditional patterns of care are changing, economic and emotional support within families is still stronger than within many advanced capitalist countries (Apt 2002).

[Re]constructing the moral economy of care in Asia

Any attempt to understand ageing and care for older people in Asia needs an appreciation of its complex cultural landscape. Asia is made up of a diverse mix of developed and developing countries. Decreasing birth rates, combined with increased life expectancy means population ageing is progressing rapidly. Indeed, countries such as Japan, Taiwan and the Republic of Korea have some of the most rapidly ageing populations in the world (UN 2007). It is worth noting, however, that in countries where Catholicism and Islam predominate (for example the Philippines and Malaysia respectively) high fertility rates continue (Glaser et al. 2006).

Whilst traditional family-based systems of care still predominate, the socio-cultural roots that underpin who cares for older people differ. In cultures where older people are considered repositories of religious teachings and beliefs, for example, among Malaysians and Singaporeans, preservation of their social status is likely to be high (Chan 2005). Conversely, in cultures where youth and economic success are increasingly seen as priorities, older adults can find themselves no longer having a social role to play, resulting in negative outcomes for their health and wellbeing (Lau and Pritchard 2001).

Familial systems of care in Asian countries may either be based on patrilineal systems that stress the responsibility of sons (and their wives) or more flexible bilateral systems where daughters play an equal (or possibly more) important role to that of sons (Chan 2005). Patrilineal systems of care are based on Confucian beliefs of filial piety, fraternal duty and the ethical concepts of respect, love and support (Kim and Maeda 2001; Zhan and Montgomery 2003). The giving of care to an older family member is thus linked to those moral obligations that tie 'the family life cycles of descendants and ancestors in their public and private worlds of material and spiritual existence, and consciousness' (Holroyd 2003, 158). In this family system, vertical family relationships are emphasised and older people's dependency on younger members is valued and encouraged (Yamato 2006). In countries such as China, Japan and the Republic of Korea, the system of filial piety means that sons, rather than daughters, have been the traditional providers of physical and financial care with daughters-in-law, rather than daughters, taking on the primary care-giving role (Kim and Maeda 2001; Zhan and Montgomery 2003; Hanaoka and Norton 2008). Even during the communist regime in China, this system of care remained in place.

Traditional cultural beliefs about whose responsibility it is to care are currently being undermined by economic change and significant shifts in the population structure across Asian countries. Rapid socio-economic development combined with rapidly ageing populations within these countries is also impacting on patterns of morbidity and mortality. There has, for example, been a growth of diseases associated with old age (such as heart disease, arthritis, stroke etc.) resulting in increases in the prevalence of chronic ill-health and functional impairment. These changes have enormous implications in terms of the form and extent of care support older people may require with attendant implications for families as the traditional suppliers of that care. In some countries, as in the Republic of Korea, the impact on traditional cultural norms of parental care-giving has been exacerbated by a relative absence of any public social services to meet the needs of older people (Chee and Levkoff 2001).

Interestingly, a recent comparative survey in Japan and the Republic of Korea noted that despite the fact that the mean age of Japanese respondents was much older that that of their Korean counterparts (84.7 years and 76.6 years respectively), older Japanese people were significantly more able than their Korean counterparts (Kim and Maeda 2001, 248). A number of contributory factors may account for this, including the increased time-span over which the ageing of the population occurred in Japan, higher education levels and better state policies to protect older people. Increases in education and income levels may also mean that children are more capable of providing financial and emotional support to parents. Chan (2005) notes, for example, that increased education amongst female children in Taiwan has significantly affected the level of financial transfers daughters make to their parents.

Structural constraints are also important in the attempt to interpret differences in the contours of care. In China, for example, the ageing of the population, combined with the implementation of strict population control through the introduction of its one-child policy in 1979, has manifest in a top-heavy population structure – one that undermines traditional informal care-giving (Zhang Wenfan 2002). The ageing of the population has also outpaced economic development in China, which combined with low GDP per capita, is making it difficult for individuals and families to cope financially, with care support. As a consequence, some researchers have pointed to a shift away from traditional systems based on filial piety toward a more bilateral system in which daughters are increasingly taking on caring responsibilities for their ageing parents (Zhan and Montgomery 2003). Whether filial or bilateral systems of care prevail, as family size decreases, adult children lacking the benefit of sibling support are finding it increasingly difficult to cope with the numbers of frail older people requiring care (Zhan and Montgomery 2003). A recent commentary in the Beijing Review (2007) went so far as to claim that the family based system of support for older people was expected to come to an end within the next 20 years. As in other developing countries, the decline in the population available to care has been exacerbated by mass migration from rural to urban areas. Whilst urban populations doubled from 16 percent to 32 percent between 1960 and 2000, three-quarters of all elderly people in China still live in rural areas (Fu Hua and Xue Di 2002, 4). Hence, as elsewhere, the migration of working age adults to urban areas is leaving behind a residual population of frail elderly and disabled people who would formerly been supported within a familial system of care (Zhang Yuanzhen 2001).

Alongside these demographic and structural changes, families in China have also struggled to survive the transition from a socialist to a free market economy. The privatisation of many enterprises previously almost wholly owned by the state has resulted in a contraction and closure of many businesses with a loss of job security and healthcare benefits – an impact that is disproportionately experienced by women and older people. Due to the greater proportion of women working in the informal sector, they are also less likely to have pensions than men, and thus are forced to turn to their children to meet their care needs. As familial systems of care are breaking down, however, increasing numbers of older people are now living alone (Zhan and Montgomery 2003). As elsewhere in Asia, this points to an urgent need to rethink how care for older people is provided. More specifically, it has been suggested that Asian countries need to move away from sole dependence of family care to focus on the development of a form of ageing in place that combines formal services with family care (Zhang Wenfan 2002). Such views are part of an ongoing debate about the role of state versus familial responsibilities to care in many Asian countries.

Despite the fact that physical and economic support has traditionally been seen as the responsibility of families, over the last few decades a number of Asian societies have begun developing state-based programmes to support older people. Indeed, most countries in Asia now have some form of state-run pension or social security programme. However, coverage varies greatly and tends to be better developed in countries with higher GDP (Chan 2005). Mehta (2005) maintains that with the exception of Japan, the present mosaic of services for family caregivers in the majority of countries in East and Southeast Asia is minimal. Not only is there a shortage of trained social workers and counsellors but there is a severe the lack of funds within many of the voluntary welfare agencies that deliver community-based services. For example, though social eldercare in China is being developed through a range of public, private and voluntary services, the development of community care services in China is still in its infancy (Hua and Xue Di 2002). Furthermore, while medical services in the home have been available for some time, this does not include personal or social care services. Though day-care, meals-on-wheels and care-call services for lone elderly are developing, they are currently few in number. As with most healthcare in China, these services are paid for either through health insurance or on a fee-for-service basis. State subsidies are limited and the relatively underdeveloped social security system further weakens families' ability to care (Kequin et al. 2000). Where the family is unable to provide informal care, personal care tasks must be undertaken by privately hired home care workers. Institutional care in the form of housing for the elderly, care homes and geriatric nursing does exist, but as elsewhere, is seen as the option of last resort. Few choose this option and those that do are mainly without relatives to take on the care-giving role (Kim and Maeda 2001). Hence, as with other systems of care traditionally rooted in familial support, care is firmly based in the private space of the home.

Given the speed at which population ageing is occurring in Southeast and East Asia and the immense social and economic changes that the region is experiencing, governments of these countries have much less time to react to these changes compared, for example, to their Western counterparts (Chan 2005). Seeking solutions to this looming care deficit is thus critical if a crisis in care for older people is to be avoided.

Care-giving in Latin American countries

A key difference between Latin America countries and more developed countries is the relationship between the speed and size of the momentum toward ageing and the social and economic contexts within which this is taking place (Wong et al. 2006). So while the ageing of society in developed countries took place long after relatively high standards of living were achieved and institutional strategies to offset the worst effects of residual inequalities implemented, Latin America is experiencing a highly compressed ageing process. This ageing of society is also occurring at a time when economies are fragile, poverty is rising and social and economic inequalities are expanding rather than contracting. Thus, not only are Latin American countries ageing 'prematurely' (Wong et al. 2006, 159), but institutional support for older people who require care is more likely to be reduced rather than increased. This 'care deficit' has been compounded by the complex socio-economic change occurring within many Latin American countries and the impact this has on both the social structure and mobility of their populations.

Whilst institutional strategies to address the care needs of older people have been developed, the extent and effectiveness of these strategies varies across the region. Further, as some commentators have noted, in many countries access to collectively financed services and resources is contracting (see for example,

Barrientos 1997; Cruz-Saco and Mesa-Lago 1998). The imperatives arising from an ageing population structure and declining economy are compounded by the fact that meeting the long-term care needs of older people – either formally or informally – is not usually seen as the core function of health care policy in Latin America. Rather there is an assumption that family and community structures will reduce the need for institutionalised care. Lloyd-Sherlock (2000) however notes that in Uruguay, at least, this assumption is open to question. Whilst nominally it is a country that places a high value on familial care, the proportion of older people living in residential care is higher than in the United States. This, however, is something of an anomaly as residential care in most other Latin American countries is private and thus beyond the reach of lower income households.

What are the implications of these issues for the care of older people across much of Latin America? Clearly the complex issues that impact on how this is played out will vary across countries. Traditionally, however, care has been embedded within family based systems of support, though in common with the experiences of other countries that have gone through a period of development, this is changing.

In Argentina, for example, older people are likely to have experienced a period of development and prosperity during their early lives on a par with developed countries. During the mid-twentieth century urban industrial centres boomed, populations migrated from rural areas resulting in the decline of the extended family and a fairly all-embracing welfare state was developed. Economic instability from the mid-1970s, however, saw the virtual collapse of its urban industry, with a consequential growth in unemployment and poverty (Wong et al. 2006). These economic shifts have had complex effects on intergenerational relationships and responsibilities for the care of older people. Whilst Argentina has a distinct system of healthcare for its older people, the system has long been characterised by bribery and political corruption. Hence the integrated medical care programme available to older people has limited credibility, efficiency or functioning (Lloyd-Sherlock 2003). More recently, Lloyd-Sherlock (2008) has pointed to a further shift in familial structures manifest in growing tendency for adult children to remain in the parental home as a result of the growing strain felt by families during periods of economic instability and social change. At least 37 percent of all households now contain at least one older person. It thus is unclear to what extent norms and attitudes to care have changed. For some commentators, the shift back towards more familial-based system is indicative of the resilience of intergenerational obligations based on religious (Roman Catholic) values (Varely and Blasco 2003). Others, however, argue that the earlier period of modernisation and economic growth has weakened these traditional familial systems and lead to a growth of youth-centred, individualistic values. As a consequence this shift should be viewed simply an outcome of economic instability and pragmatism on the part of adult children (Lloyd-Sherlock and Locke 2008).

Conversely, though a relatively small, lower middle-income state, Costa Rica has a highly developed welfare state. Approximately two-thirds of its population have access to a pension scheme, with alternative means-tested state benefits available for those not covered by state of other pension schemes. The rights of older people with regard to housing, labour conditions and well-being are also enshrined in its 'Integral Law for Elderly People' (Knaul et al. 2002). The law further defines the duties of state and health services and regulates private care institutions to ensure the rights of older people are maintained. Male life expectancy exceeds that of more developed countries – including the Unites States, Germany and Finland. It also has a relatively young population in comparison to other more developed countries with only 7.3 percent of its population over the age of 60 (Knaul et al. 2002, 39). The relative success of Costa Rica's extensive healthcare system is due to high levels of public expenditure on public institutions specifically designed to alleviate poverty and the maintenance of a highly centralised health and welfare system that goes against current global trends toward the marketisation of healthcare. The majority of its population are also concentrated in metropolitan areas, maximising service coverage. Despite these developments, care-giving for its frail older populations is still considered a family responsibility affecting the provision of long-term care services. The relatively low levels of women in the workforce (though increasing) means there is still a wide availability of women to undertake family care and support for frail older people. Daughters and daughters-in-law in particular take on most of the care-giving tasks for their older relatives. Family care provision is changing, however, due to shifts in living arrangement and the rise of the nuclear family. The state has sought to address this by taking on the responsibility for integrating the family into the care of older relatives through training programmes for informal care-givers (often undertaken by NGOs and volunteer programmes).

Until the mid-1990s, residential care services were not considered an important alternative form of care for older people in Costa Rica. Formal care was provided through specialised hospitals and targeted programmes. Furthermore, the State healthcare system strongly discouraged the growth of private provision (Knaul et al. 2000). Despite this, Lizano (2000) notes that there are three forms of privately managed care support for older people in Costa Rica: sheltered housing, nursing homes, and day-care centres. To qualify for either of the first two, however, an older person must be without familial resources or care. Those older people unable to make use of day-care services have access to a programme of community care services. The extent to which Costa Rica may be able to maintain this level of commitment as its elderly population grows, however, will clearly be a matter for concern.

Similar concerns about how to maintain its ageing population are growing in Brazil, where the health and welfare system is poorly equipped to deal with the problems of an ageing population. What was once viewed as a generous public pension scheme, for example, is now being seen as an economic liability. At present, there is a marked lack of proper care services for frail older people, particularly intermediate care services such as day-centres, day hospitals, community centres and so forth. Those who have no family or basic survival conditions live in institutional settings, but the Brazilian culture is prejudiced against committing frail older people to care homes, hence those that do exist are few (though growing) and generally of poor quality. Though policies developed in the 1990s aimed to enforce an improvement in standards, to date they have been largely ineffective. Ramos (2000) maintains that given the fact that demand for places exceeds supply; there is little incentive to effect improvements. The lack of social support and residential care places the main emphasis on family care within the home. Many middle-aged women thus find themselves maintaining a job, caring for their own children and caring for their frail elders. The outcome, as one respondent in a study by Machado (2001, 11) commented, is that 'within their own family, [the elderly – sic] are more and more excluded from society, more at home with no support, no-one has time for the elderly, everybody goes to work.'

Commentators maintain that while Brazil's most pressing problem had been its growing army of glue-sniffing street children the problem has now shifted with its ageing population now being its most important challenge (Ramos 2000). Longman et al. (1999) go so far as to claim that if the government fails take urgent action, Brazil may be faced with a future problem of street *elders* without having solved its problem of the street children. Yet according to Ramos (2000), despite the escalating problem of care for older people, the Brazilian government still see it as less critical than the problem of its street children.

What has contributed to the rising problem of caring for an ageing population in Latin American countries? In part, the pattern of development closely resembles that of other lower middle income states. That is, it has been characterised by rapid industrialisation and urbanisation from the mid-twentieth century, declining birth rates and a shift in morbidity and mortality patterns away from infectious disease to the diseases and chronic health problems of old age. These factors have all acted to undermine the pre-eminence of the family care-givers to their frail elderly relatives.

A number of other factors have also contributed to our understanding of who cares and where in Latin America:

- First, there are significant regional variations in life expectancy. For example, Ramos (2000) notes that there is over 15 years difference in life expectancy rates between the poorest north eastern areas and the wealthy southern areas of Brazil. Similarly, birth rates vary, with women in the poor north east averaging families of five children and women in the south achieving only replacement birth rates. On the face of it, this would appear to indicate that older people in these poor rural areas are more likely to be able to draw on familial care. In reality, however, many young rural dwellers of working age migrate to urban areas leaving the rural elderly with no familial support, contributing to the problem of ageing rural migrants with no familial networks.

- Second, many Latin American counties have traditionally been patriarchal societies – an ideology that has been embedded in family structures and national policies until relatively recently (Goldani 1990). Women's work was conceived as supplemental to the male breadwinner, justified only where the head of the family could not afford to support the family. Until relatively recently, therefore, women's participation in the labour force has been fairly low – and regulated in some countries. While more liberal legislation began to emerge during the 1970s and 1980s, there were still prejudicial barriers to women working. Goldani (1990) maintained that it was only with the introduction of a constitution in Brazil 1988, for example, that the pattern of the patriarchal family was abandoned. Since then, there has been a growth of childless couples, one-parent and female headed families. Such changes have an inevitable effect on the status of older people and women's ability to care.
- Third, familial responsibility for older people requiring care has traditionally been unquestioned, particularly amongst the poorer populations. Rapid urbanisation combined with low wages, high rates of accident due to occupational hazards, significant housing shortages in urban areas and geographical distance, however, have all acted to undermine adult children's ability and willingness to care. Family and kin networks are losing ground before societal mechanisms to effect institutional transfers are securely in place (Ramos 2000; Wong et al. 2006; Lloyd-Sherlock and Locke 2008). Where family care is not forthcoming and residential care is unaffordable or stigmatised, the only option may be hospitalisation – a widespread problem in many developing countries, and one which is likely to get worse (Lloyd-Sherlock 2000).

Post-socialist countries

Finally, it is worth considering some of the factors that have affected how care, and who cares, has been [re]constructed in the post-socialist states of Eastern Europe. Within countries formerly part of the Soviet Republic, the historical legacy of collective responsibility meant that until relatively recently, (basic) long-term care for frail and vulnerable groups was provided within state-run institutional settings. The ideological commitment to collective provision, together with high levels of women in the workforce, reinforced state commitment to the development of institutional care rather than individual community-based supports. The legacy of socialism in these countries, thus, created a very different ethos of care (Bezrukov 2002). Though experiences vary, by and large the contours of care are neither typified by a culture of familial obligation nor are they comparable with those of western democracies where the concept of 'ageing in place' and the development of community care services have increasingly taken root since (at least) the mid-1970s.

The post-socialist states of Eastern Europe have experienced significant political and structural transitions since the end of the twentieth century. As Najafizadeh (2003) points out, when the state no longer assumes the core role as the provider of care, who will assume that role becomes a critical question. Perceptions of whose responsibility it is to care for older people and where that care should take place are shifting. The socialist history of the Ukraine, for example, means that while the healthcare system has been widely available across the country, the availability of key services has been limited. Though community care and home based supports for older people are developing in post-socialist Ukraine, they are relatively new.

The Ukraine, however, has a large elderly population – currently 20.5 percent of the population are over 60 years of age – a figure set to rise to 25 percent by 2025 (Bezrukov 2002, 1). Its ageing population means that the size of the population unable to work is beginning to exceed those who can. The capacity of its informal care system is thus very limited and as such cannot form the dominant model of care for older people. While state legislation gives all citizens the right to healthcare and medical aid – either through state financial support, or free of charge at state/community institutions – the gap between budgetary allocations and real cost means that some services are either partially or fully paid for on a fee-for-service basis (though specified population groups received fully or partially subsidised care for all services).

The incapacity of its informal care system means that alternative public or privately provided systems need to be developed. Some voluntary sector (or NGO) provision is beginning to emerge – including geriatric hospitals and homecare for frail older people. Though the development of an organised voluntary sector was not encouraged under the former socialist regime, there has been a fairly long-standing tradition in the Ukraine of rendering assistance to disabled and frail older people on a voluntary basis. This tradition has grown in recent years with volunteers working in a range of care services including social services, welfare, in-patient and medical centres as well as within formal voluntary organisations themselves. Parishioners from differing religious groups also play a role in providing homecare to frail older people. Overall, as Verzhikovskaya et al. (1999) note, there is no one single system of long-term care provision for older people. Most is supplied by the state, mainly through local budgets, on the initiative of local administrators or other potential funders (e.g. religious or voluntary groups) with priority being given to those who live alone. Differences in resources and infrastructure at regional and locality level mean that frail older people and their informal carers can have highly differentiated access to formal care services.

Since the collapse of the Soviet welfare state, poverty has affected an increasing number of older people. Collective healthcare systems have become less reliable and social care for older people has almost entirely ceased to function. This is particularly problematic in rural areas. Whilst a pension system exists in Russia, its value has dropped to just above subsistence level. The insecurity of personal welfare has been a persistent feature of the history of the Russian people, indeed, Tchernina and Tchernin (2002, 560) note that 'personal responsibility for one's own survival is ancestral and quasi-perpetual'. A distinctive feature of Soviet and post-Soviet society in Russia has been the strength and distinctive role of informal social networks. Vishnevskij (1995) maintains that these are the residue of forced urbanisation in the 1930s, resulting in the creation of urban societies that retained those rural practices that had characterised Russian rural communities for centuries. During the Soviet era older people developed a range of survival strategies commonly based on these informal social networks and the development of a range of income generating activities. Such activities included barter and exchange in the grey and black economies, street trading of foodstuffs and other goods, trafficking of illegal goods and begging (Tchernina and Tchernin 2002). Many young adults, unable to obtain their own apartment or home spent the first ten years or so of their marriage living with their families. Younger adults would take on a breadwinner role whilst older people took on household and child-rearing tasks, garden cultivation etc. From the 1990s, however, there has been an inversion of the breadwinner role with many older people drawing on their ownership of their own apartments, garages and allotments to generate income that they frequently redistribute amongst their children's families. Most expect neither reciprocal exchange nor financial support from their families (Tchernina and Tchernin 2002). Alongside the decreasing responsibility of the state for its older people there has been a growing practice of legal agreements between older people and entrepreneurial relatives who pay a monthly sum (equivalent to a market rent) for the use of the older person's apartment plus any personal services rendered. These developments highlight not only how older people are increasingly taking on responsibilities to care for their families rather than vice versa, but also the frailty of intergenerational solidarity.

It is important to note, however, that the experience of post-socialist countries that were not part of the former Soviet Union can be very different. In the old socialist regime in Albania, for example, care was characterised by a dual system of family and state provision in which patrilocal systems of care (in which the youngest son took responsibility for parental care) still prevailed. The sudden collapse of Albania's highly authoritarian regime in 1989, however, lead not only to a collapse of its paternalistic state welfare system, but a massive backlash against the old regime's tight regulation of movement – both within and outside Albania. Internally, this regulation had acted to keep urbanisation artificially low (Vullnetari and King 2008). Hence, once the barriers to movement came down Albania experienced massive population mobility driven by the need to survive the economic and political chaos that followed in the wake of the collapse of its socialist regime. The Albanian Government estimates that since the movement ban was lifted, around one million people – that is 1:4 of those aged between 20 and 40 years of age – have migrated, mainly to Greece and Italy, but also to North America, the UK, France, Belgium and Germany (INSTAT 2004). This scale of migration, concentrated within such a short time period is unprecedented in modern Europe and whilst many migrants still send remissions back to their ageing parents, it has resulted in a severe breakdown in trans-generational care.

The combined collapse of state welfare together with the 'care drain' created by the mass migration of adult children has produced a new phenomenon of abandoned and destitute older people. Vullnetari and King (2008) maintain that only five residential care homes for older people remain in Albania together with a handful of NGO run day-centres – all of which are located in major cities and are clearly insufficient to cope with the needs of an urbanising population. Social care for older people in need of support is virtually non-existent in rural areas. Yet while villages have next to no public services, Vullnetari and King (2008) point out that the outflow from rural areas has left no shortage of housing for older people. Combined with the sense of support and community that is still strong in these areas, this to some extent is helping to ameliorate the effects of the 'care drain' and relieve pressures that would otherwise be exerted on state welfare services.

Concluding Comments

In considering the landscape of care from an international perspective it is possible to see that how care for older people is manifest is shaped by a range of differing cultural, political and economic constructions that change over time and place. All of these impact not only *who* cares, but *where* that care takes place. Rights and responsibilities in relation to the care of older people operate across a spectrum; with care being located firmly within the family and home at one end of the spectrum, and within the state and institutional settings at the other. In between these two extremes a number of differing systems of care exist that incorporate a range of providers from informal care-givers to state, private and non-profit actors. These actors also operate across a range of places including domestic, community and institutional settings. How care is produced in differing places, however, is also affected by a range of factors that operate not only at the national level, but also at regional, local and community levels. The construction of care is also shaped by the prevailing political ideology and how care systems are organised within that ideology. Though this is manifest largely at the level of national government, regional and local government can also play a significant role here is shaping local landscapes of care.

Also of significance is the historical legacy of care and the impact of wider global processes. The ageing of the world's population and changes in the world's economy, for example, are having a significant impact on the extent to which both governments and individuals are able to maintain existing patterns of care for the growing numbers of older people. As this chapter has demonstrated, this is not simply a phenomenon of advanced capitalist states. The complex interplay between ageing populations, modernisation, urban growth and migration patterns, together with changes in the composition of family and workforce patterns as well as structural shifts are all interacting to create a complex web of factors that affect traditional patterns of care with, as yet, little by way of replacement systems of care. Elsewhere, the collapse of previous political and/or economic regimes (as in post-socialist and some Latin American states, but also to some extent in traditional welfare states) has left significant gaps in care formerly offered by the state that families are either unable or unwilling to fill – the so-called 'care gap'. Addressing the issue of how we care for our older populations in the twenty-first century is thus an issue for developed and developing countries alike.

Finally, it is important to note that whilst we are experiencing global ageing and an increased likelihood that greater numbers of older people will require care, they can also (and often do) make an important *reciprocal* contribution to care. Within familial settings they may be net contributors to care, either economically, through the pooling of financial resources or other assets (such as the home, land or a family business) or the undertaking of childcare and other household tasks. These important contributions can also act to offset the worst effects of social isolation amongst older people. Hence, it is important that we do not disregard the two-way nature of care, or the value of recognising this reciprocal care, particularly within familial relationships.

Chapter 5
Care and Home

As noted in Chapter 4, globally not only are there are marked variations in the balance of resources invested in care for frail older people, but these variations determine who cares and the locus of that care. Tucker et al. (2008) suggest, however, that with regard to many western societies, we have begun to see evidence of a convergence in care policies manifest in a greater concern for supporting older people to remain within the home. The rationale for this shift, they maintain, is not only a belief that home is the preferred site of care, but that it is also more cost effective. So while this chapter focuses largely on how older people and their informal care-givers experience ageing in place in the UK, the shift toward convergence means that many of the issues raised in the chapter will be of relevance to other western societies.

Framed by debates about the ageing of society and its implications for care in the UK, the chapter discusses how older people and their informal carers experience care-giving in both home and community settings. In doing so, it considers how they understand and gain access to public, private and voluntary care services and the importance of place in mediating the availability of, and access to, these services. The chapter also considers the meaning of home for older people and their informal carers. In doing so, it considers how, as levels of care needs increase, this can result in changes in feelings of privacy, control and power relations between formal and informal care-givers and a blurring of the boundaries between institution and home.

Contextualising Care and Ageing in Place in the UK

It is not the purpose of this chapter to chart the historical development of care for older people in the UK – this has been well documented elsewhere. Nevertheless, to understand the implications of ageing in place for both older people and informal care-givers, it is important to have some appreciation of the policy context that has contributed to the shaping of these experiences.

One might, of course, argue that at its core, care for older people in the UK has always been the responsibility of friends and families with institutions such as the state and church playing only a residual role for those most frail and needy with no other visible means of support (e.g. through parish relief or institutional 'care' with poorhouses). Indeed, if we accept Offer's (1999) argument, the state's acceptance of responsibility for the care and support of older people post 1945 should be seen as the exception rather than the rule. This said, it is against this period that most contemporary changes in care and support for frail older populations are measured and framed. Of most significance has been the ideological and political shift in the latter half of the twentieth century away from institutional toward community-based care. This shift has resulted in a whole raft of legislation, guidance documents and reports that have shaped not only how, but where care and support to older people should be provided. Key changes affecting the shape of care for both older people and their informal carers, from the 1970s onwards, are summarised below.

As part of the wider shift in thinking about how best to care for frail and vulnerable groups in our society, a series of government papers and reports emerged during the 1970s and 1980s. These culminated in the 1990 NHS and Community Care Act, which pointed to the need to develop better domiciliary, respite and community-based initiatives to enhance the ability of these individuals to remain in community settings. For older people, this marked a shift from away from the earlier focus on care within large (mainly public sector) residential settings as the only real alternative to family care. Such settings had increasingly come under heavy criticism for the poor quality of care offered and their institutionalising effects (see Townsend 1962 1965). Responsibility for developing and implementing community-base services was placed in the hands of Local Government, with the National Health Service (NHS) responsible only for the medical and nursing aspects of community-based care. Local Government was required to develop a competitive market for social care through the encouragement of independent sector provision. The voluntary and private sectors were, thus, to be encouraged to compete for contracts to develop and deliver care services to older people and their informal carers. Local Government was to draw back from direct provision in favour of a monitoring, commissioning and contracting role. As a consequence, the UK saw the voluntary and private sectors begin to develop as key providers in the 'care industry'. The voluntary sector in particular began to emerge as a significant provider of personal and social care and support focused on the home and community, whilst the private sector invested heavily in residential care homes (encouraged by the substantial subsidies available for means-tested places). This is not to say that these two sectors focused solely on these distinct areas of care, but it is certainly true that the majority of care homes in the UK are in the hands of the private sector. The Commission for Social Care Inspectorate Report (CSCI) notes, for example, that over 75 percent of residential care homes in England are privately owned (2009, 22). Most public sector homes have now closed or been taken over, and while the voluntary sector does have some stake in residential care, many of these homes (though not all) are underpinned by a particular religious or cultural ethos. Similarly, the voluntary sector is the major provider of non public sector personal care and community-based services. While private sector involvement in home-based nursing and personal care is growing, during the late 1980s and 1990s, its main area of growth in terms of care provision for older people was in the residential sector.

The growth in independent sector care, however, has manifest in substantial inequities in access to, and the availability of, care services – both regionally and within specific geographic localities. In part this arose as a consequence of local political disagreements about whose role it was to provide care, and hence differences in Local Governments' willingness to wholeheartedly embrace local state withdrawal from care provision; in part as a consequence of differential decision-making between and across all three sectors about what services to develop where; and in part as a result of the long-standing separation of responsibilities for social and nursing/medical care between Local Government and the NHS that stretches back to the inception of the welfare state. Disagreements between health and social care providers were particularly evident around those services deemed to be at the cusp of medical/nursing and social care. In part this was linked to 'turf protectionism' and in part to disagreements around whose budget these services would be attributable to. This has been epitomised in the somewhat heated debate that emerged around the 'social' versus the 'medical' bath (see for example, Twigg 1997; Griffiths 1998; Lewis 2001).

Variations in care provision were further exacerbated by a continued lack of coterminosity between Local Government and NHS boundaries. I have discussed these issues and their outcomes in terms of geographical inequities in statutory and independent sector care for older people and informal carers in detail elsewhere (see Milligan 2000 2001; Milligan and Conradson 2006). While I do not propose to restate these issues here, it is worth noting that current policy is still actively seeking to redress these problems through the development of more integrated care services. Indeed, Wanless (2006) points out that while many developments of the early 1990s were deemed controversial at the time, retrospectively, it is striking to realise how similar these issues and agenda were to those still facing policy makers today. Not only does this highlight how challenging the community care agenda has proven to be, but also the relative failure of services to achieve this vision over the past 20 or so years.

The change of Government from Conservative to Labour in 1997 did not mark any substantial shift in the direction of social care. Indeed a White paper published in 1998 explicitly reaffirmed the Government's commitment to community-based care – not through any shift back toward public sector provision, but through the promotion of the so-called 'third way'. This approach explicitly rejected both the previous Government's commitment to the privatisation of care and the 'one size fits all' model of universal healthcare that characterised 'old style' Labour provision (Wanless 2006). The 'third way' had at its heart a commitment to a rejuvenated and improved partnership working designed to promote so-called 'joined-up' working. Local strategic partnerships (LSPs) around geographical or thematic areas were to bring together a whole range of actors from relevant statutory, voluntary and private sector services to address a common goal (Milligan et al. 2007). This joined-up approach was a defining feature of the 1998 initiative 'Better Government for Older people' – a pilot designed to develop local partnerships whose task was to develop strategies to meet the needs of their older populations. The complexity inherent within these partnerships, however, has meant that lines of responsibility have often been unclear, undermining their effectiveness (Wanless 2006).

Tanner (2001) has further argued that despite the rhetoric of a needs-led community care policy, this has not been sustained in the face of reduced budgets and increased demand for social care. Rather what has emerged has been a targeting of services rather than the delivery of services based on an assessment of need. This need to contain resources has manifest in a process dominated by managerial rather than professional concerns. The overall result has been an intensification of services focused on high level needs that leave limited resources for preventative care. So while the 2001 National Service Framework (NSF) for Older People and the 2006 White Paper *Our Health, Our Care, Our Say* both outlined reforms designed to firstly, improve the quality of health and social care support for older people; and secondly shift away from intensification toward the development of more preventative measures, these services remain patchy and unco-ordinated. Indeed, Tanner maintains that though the NHS has sought to focus on intermediate care as outlined in the NSF, in effect what has emerged is a focus on the development of *alternative* forms of provision for those who already meet the eligibility criteria. Hence, not only is there a continuation in the intensification of care, but as the CSCI (2009) points out, in England, there is an ongoing geographical inequity in the quality and availability of the different types of services that are available. Interestingly the CSCI also note that this variation is not just geographical, it also exists in the *quality* of care between sectors – with voluntary providers significantly outperforming private providers of domiciliary and residential care services (CSCI 2009, 23-26).

It is perhaps pertinent at this point to summarise the financial landscape of care support in the UK. All medical/nursing care is free at point of contact though personal and social care (with the exception of Scotland) is subject to means-testing. Medical and nursing care is the remit of the NHS and is paid for through general taxation. Personal and social care comes within the remit of Local Government who target services at those most in need and means-test those individuals assessed as qualifying for services. Funding for the subsidised elements of these services comes in part from central government budget allocation and means-tested benefits; and in part from local taxation payable by all householders and businesses within the jurisdiction of each Local Government. Whilst an individual can choose to enter residential care and pay for it privately, Local Government also subsidises places within care homes and placement here is also subject to means testing. Since 2002 any nursing care received within the care home is deemed to be free and as with other health care is paid for through central taxation. The 1998 Royal Commission on Long-term Care recommended that personal care be provided free to older people in the UK. While Scotland chose to implement this recommendation, England, Wales and Northern Ireland chose not to. Hence, contrary to the rest of the UK, personal care in both domiciliary and residential care settings in Scotland has been provided free of charge at point of receipt since 2002 and is instead paid for through central budgets (Bell and Bowes 2006). Debate continues within the rest of the UK about how future social care needs should be developed and funded – an issue that is currently the subject of a major government review (HM Government 2008).

One further development of note has been the introduction of the Direct Payments option referred to in Chapter 2. Initially introduced in 1996, the Act introduced powers to enable certain categories of people to be offered a cash sum in lieu of services which they could then use to purchase their own support (Wanless 2006). In effect, this makes individuals in receipt of Direct Payments employers, able to make choices about what forms of care and support they wish to purchase. This option has proven popular with the disability movement and has been seen as a critical move forward in its bid to replace the notion that services should be framed around care to one of support (Thomas 2007). The Direct Payment option was extended to older people in 2000, but uptake was limited, hence in 2003, the Government extended the scope of direct payments to make it a duty, rather than a power, for this option to be offered to eligible people. Though uptake amongst older people is now rising (CSCI 2008), growth remains slow – despite this being the largest group that could potentially benefit (Argyle 2004; Manthorpe and Iliffe 2005; Swift 2007). Several reasons have been put forward for this; in part it has been linked to confusion and a lack of knowledge and clarity about the role of health services in promoting direct payments; in part about the unwillingness of older people and their informal carers to take on any additional stress and responsibility that may be involved in becoming an employer; and in part about a lack of choice over services offered within their locality. This has been compounded by confusion amongst older people and their informal carers about what services are actually available and how to access them. Indeed, in a recent study of the future of long-term care for older people in London, Robinson and Banks (2005) were lead to conclude that not only was the so-called care market failing older people, but efforts to empower them, and their carers, through direct payments were doomed to failure unless sufficient services of the kind that people actually wanted to purchase were developed.

Informal care and ageing in place

Despite the shifts referred to above, the implementation of policies designed to support ageing in place are still heavily dependent on informal care. Government recognition of this is evident in a range of policies that have emerged particularly from the mid-1990s, designed to support and encourage informal carers to continue caring for as long as possible. In 1995, for example, we saw the implementation of The Carers (Recognition and Services) Act – with subsequent Acts in 2000 and 2004 that articulated more clearly the rights of carers and the responsibilities of Local Authorities. For the first time, those who were providing substantial levels of informal care and support had the right to an assessment of their own needs over and above those of the person they cared for.

Although a pessimistic view would see this shift as having more to with preventing carer breakdown (and hence the need for state intervention) than any real recognition of the significant role informal carers played in the care of their frail older relatives, these shifts do mark the beginnings of a recognition that informal carers are not solely a resource against which other services should be measured, but that they are also co-workers that may potentially become co-clients. As well as a range of (admittedly limited) state benefits, carer support has been designed around a range of domiciliary and residential respite care, counselling, advocacy, support groups, information and training. Current provision of services to support informal carers, however, varies significantly with location and personal circumstances – for example, co- or extra-residential status. It has also been argued widely to be inadequate (see for example, Milligan 2001; Keeley and Clarke 2002; Maher and Green 2002; Arksey and Hirst 2005; Wanless 2006).

The most recent Government report, *Carers at the Heart of the 21st Century* (2008) sets out a new ten year strategy for supporting informal carers. In doing so, it recognises the dichotomy presented by the promotion of policies designed to support ageing in place – with their reliance on continued informal care-giving – and the impact of changing family structures, work-life patterns and expectations. As elsewhere, such shifts are impacting on people's willingness and ability to care, leaving some older people isolated with little or no support from family or friends. The strategy specifically notes that:

> These changes mean that the needs of carers must, over the next 10 years, be elevated to the centre of family policy and receive the recognition and status they deserve

and that

> the next decade must lead to major and substantial change in the everyday lives of carers and the family members and friends they support (2008, 8-9).

To address these issues, the strategy sets out a five point vision through which, by 2018 carers will:

- be acknowledged as expert partners in care and have access to the services they need to support them in their caring role;
- be able to maintain a life of their own alongside their caring role;
- have access to sufficient support so that they will not find themselves facing financial hardship as a result of caring;
- be treated with dignity and have sufficient support to ensure they maintain their physical and mental wellbeing;
- have protection from inappropriate caring (e.g. children and young people).

It would be hard to disagree with such a laudable vision. How exactly this is to be successfully implemented, however, without first tackling the ongoing and enduring problem of the health and social care divide is more difficult to foresee.

In sum then, five core themes underpin the contemporary context of ageing in place in the UK:

- a shift in the model of care for older people from institutional to community and domiciliary settings;
- a widening of care providers that includes not just the family and the state but also voluntary and private sector provision;
- enduring problems and continuing efforts to resolve the long-standing health and social care divide;
- a shift of focus from marginal services targeted at a few high-end users to more mainstream provision with an increased focus on preventative services; and
- a shift of emphasis from informal carers as resource to recognising carers as both co-workers and co-clients.

This shifting landscape of care thus represents both a policy ideal and a complex process of interaction between older people, those providing care and support and where that care and support takes place.

Care and Home

Any attempt to understand the implications and experience of ageing in place for older people and their informal carers, brings into focus the complexity of home as both a site of social interaction and personal meaning and as a site of care that brings both the public and private into tension.

The meaning of home, self and identity

Ageing in place is built on the premise that the home is the optimum space in which to provide care and support for older people in a way that will enable them to remain as independent as possible, for as long as possible. The home is seen as a setting that is both familiar and imbued with particular meaning. Implicit within this notion is an assumption that the ongoing and temporal process of inhabiting a familiar place somehow results in the development of a unique sense of attachment that is both supportive and adaptive – particularly for older people. There are at least three core aspects that contribute to this:

Home as haven Critical to understanding this sense of attachment to place is the notion of the home as haven. As a 'protected place' the home represents a site of security, familiarity and nurture (Tuan 2004) – a place of retreat into a private world, away from public scrutiny, where the individual can control decisions about who to admit or exclude. This can be particularly significant for older people who feel vulnerable outside the bounds of their own private space and where the surroundings of the home provide an important buttress to their sense of self (Sacco and Reza Nakhaie 2001; Milligan et al. 2004). The presence of private possessions and familiar objects within the home can reinforce this sense of self and social status, endowing the home with personal meaning (Rubinstein 1989). This can be particularly important for those whose cognitive and sensory abilities may be failing. Familiarity with the organisation of the home can, thus, provide an important reinforcement to the sense of self and safety. As a site of ontological security, the home becomes a familiar and 'safe space' from the threats of the outside world, so extending the individual's ability to successfully age in place. Critical to this aspect of attachment to place is the view that older people who have a positive attachment to home are more likely to feel in control, secure, and have a positive sense of self – this in turn 'helps adjustment to the contingencies of ageing and enhancing well-being' (Wiles 2009, 665).

Home as a preconscious sense of setting   The familiarity of home has also been argued by Rowles (1993, 66) to facilitate a 'preconscious sense of the setting'. For example, he points to the way in which, over time, we develop not only a physical attachment to the home, but also the routines we perform within it. This, then, enhances our ability to instinctively negotiate spaces within the home without coming to any harm and hence it is an important contributory factor in promoting successful ageing in place. This is most clearly demonstrated when the process breaks down; for example, when an individual trips or knocks over an object in the home that had never previously presented an obstacle despite being in the same location for a long time.

However, whilst this preconscious sense of setting may enable individuals to transcend apparent physiological and sensory limitations, it may also make them more vulnerable to changes in the physical environment. A small object left on floor or the moving of a small item of furniture (such as a chair or side table etc.) to a different location can, for example, contribute to a fall or small accident (Rowles 1993). Larger changes in the physical environment arising as a consequence of the need to install adaptations or rearrange the home to accommodate the needs of care-giving are thus likely to decrease the preconscious sense of setting and increase vulnerability. These changes may be particularly difficult to negotiate for those experiencing sight impairment or short-term memory loss such as in dementia.

As people age in place, not only are the spaces they inhabit likely to undergo physical change, but their sense of place and their associations with those places will change (Wiles 2005). It is important, then, to recognise that home is not some kind of static, physical structure but is a dynamic, negotiated and contested space in which understandings of home and the complex relationships between care-recipients, formal and informal care-givers are played out. Yet this can also be framed alongside images of the stresses experienced by older people and their informal carers that arise from the need to relocate to progressively more supportive environments as frailty increases (Milligan 2006). Indeed, commentators such as Danermark and Ekstrom (1990) maintain that where older people are relocated involuntarily, this can result in increased morbidity and mortality.

Home as a site of embodiment and identity   The home can also be viewed as the embodiment of identity and self-expression that not only anchors people within a particular locality, but which is manifest as a site of memories and a daily reminder of continuity with past identity and relationships. Here, Augé's (1995) conceptualisation of 'anthropological place', as a place of connection, memory and identity is of particular significance in facilitating our understanding of the importance of place in the caring experience. By conceptualising the home as a site containing shared memories and a family history in which the informal carer and care-recipient can connect and identify, the home facilitates an understanding of the needs and desires of the care-recipient. De Certeau et al. (1998, 145) further note that, 'a place inhabited by the same person for a certain duration draws a portrait that resembles this person based on objects (present or absent) and the habits they imply'. Exclusions and preferences, order and disorder, the organisation of available space, colour, texture and use of materials and so forth. all serve to compose a 'life narrative' about those living in the domestic space. Effectively harnessed, formal care-givers can gain important clues about the identity and personality of the care-recipient that can act to enhance the caring relationship. As the embodiment of identity, the home can act to place limits on the extent to which an individual can be objectified and depersonalised as within a collective [institutional] regime.

For de Certeau et al. (1998, 145), the home represents a personalised, private territory where 'ways of operating' are invented, deployed and repeated from day to day in ways that gain a defining value. As sites in which the familial microcosm is rooted, such spaces are both materially and emotionally dense. On the one hand they can be construed as 'protected places' where the sick body can find refuge and care away from the pressures of the social (collective) body; on the other, they represent 'theatres of operation' in which a multiplicity of functions and practices take place. This spatial dialectic not only reinforces the porosity between public and private within the home but requires us to recognise and unpack the shifting power-dynamics of care that occur within these places.

As Bamford and Bruce (2000) point out, maintaining a sense of control and autonomy in the face of increasing and unwelcome dependency is important for both the care-recipient and the informal carer. Remaining at home is often seen as an important symbol of that control and is particularly important with regard to health professionals. As one care-recipient within their study put it: 'There's a lot isn't there, to be said to come home and lock our door and it's your own place … And do what you like' (556). Within de Certeau et al.'s conceptualisation, those residing within the private space of the home have the power to exclude. Hence, unless explicitly and freely invited to enter, every visitor can be construed as an intruder. Visitors to the home enter on a privileged basis that is bound by the norms of being a guest – and guests should not trespass outside of those areas into which they have been specifically invited. Twigg (2000) and others have thus been lead to argue that as guests, formal carers require permission to enter and undertake care work, empowering care-recipients to the extent that they can both determine which areas of the home may be accessed by the formal carer, and can, if they so choose, refuse them entry. The social norms construed within this conceptualisation of home are thus viewed as reinforcing the capacity of care-recipients to say 'no' to formal carers seeking to enter the home to deliver care interventions.

Whilst this may be true, we should not forget that care practices designed to enable older people to remain within these 'theatres of operation' for as long as possible mean that the ability to exclude from even the bedroom, usually the most protected of private spaces, will be limited as levels of frailty increase. A growing requirement for support with activities of daily life such as dressing, bathing and getting ready for bed, for example, may be critical to facilitating a frail older person's ability to manage their daily lives. At the same time, however, for this care to be performed the older person must allow relative strangers to have access to some of the most private spaces of his or her home. Whilst formal carers still require permission to enter these spaces, refusal will not only result in a reduction of the older persons' ability to successfully age in place, but is likely to hasten entry to long-term residential care. Hence, those in receipt of care within the home can find themselves positioned within altered and shifting relational patterns that are instrumental in the construction, dismantling and reconstruction of their experiences of home (Angus et al. 2005).

The growing implementation of new and remote care technologies to support ageing in place is adding yet a further layer of complexity to our attempts to understand how these power relationships are constructed and played out within the home. This is discussed in more detail in Chapter 6. Yet these complexities can also be constructed in stark contrast to the erosion of an individual's power to exclude as they move into supported and residential care settings.

So while issues of meaning, identity and sense of setting are important in helping us to understand why the home can facilitate successful ageing in place, it needs to be recognised that this will shift and change with increasing levels of frailty. The critical question then becomes whether home still remains the best site of care for older people. Work around end of life care and 'place of death' (see for example, Brown 2003; Gott et al. 2004) suggests that whilst older people may initially prefer to be cared for by informal carers within the home, contrary to expectations, as levels of care needs intensify, the nature of home changes such that many would prefer to be cared for elsewhere. As Brown put it, 'home must change when people are hospiced there. So the paradox emerges: is it still home?' (841).

Home and care work

Earlier chapters have drawn attention not only to shifts in who cares, but where that care takes place. While the domestic home is now the preferred setting for the care of older people, the relocation of care work from institutional to domestic settings creates tensions between home and work that can fundamentally challenge the meaning of home. Formal care workers entering the home need workspaces that are clean, hygienic and efficient for the purpose of delivering care (McKeever 2001).

This frequently requires the reorganisation of domestic space to accommodate the 'paraphernalia of care' – and in some cases the physical modification of the infrastructure to support access (for example, through the installations of ramps, stairlifts, hoists etc.). Whilst such artefacts are routinely available in institutional settings, these spaces can be organised to conceal some of the more disconcerting features in ways that cannot be easily achieved in domestic settings (Roberts and Mort 2009). Though debate around institutional settings has highlighted how efforts have focused on manipulating the institution to make it seem more home-like (Peace et al. 1997; Milligan 2006), there has been only limited discussion of the ways in which the importation of artefacts associated with institutional care can affect the meaning attached to the domestic home (though see Milligan 2000; Gott et al. 2004). Yet older people and their informal carers are unlikely to welcome the reordering of the home as a clinical work space, instead placing value on the home as a private, comfortable and aesthetically pleasing space that is imbued with personal memories and a sense of history and belonging. So while professional care supports within the home may be beneficial to the informal carer and care-recipient, they also transgress the social space of the home, creating a change in the meaning and sense of home. As one informal carer put it:

> I felt they [statutory services] were going to send this person in, send that person in – different things, and my home wouldn't have been my own! There would have been somebody coming in all the time, every hour of the night and day, and I just couldn't stand that. My home wouldn't have been my own! (Milligan 2000, 173).

Others point to the ways in which the importation of the paraphernalia of care can reshape both the role and meaning of particular spaces within the home, such as the living room, dining room and so forth (e.g. Twigg 2000; Milligan 2001). Importing institutional artefacts is also symbolic of the physical and emotional labour entailed in caring. So when a care-recipient moves into residential care or dies, the remaining equipment can become a potent reminder of that transition. Exley and Allen (2007) and others point out, however, that informal carers can experience significant difficulty in arranging for that equipment to be removed. Indeed, informal carers have pointed to surplus equipment ranging from hoists, wheelchairs, bath aids etc. that care services have failed to remove despite numerous attempts to arrange for their pick-up (Milligan 2001).

It is important to point out however, that for some, the re-ordering of the home into a space of care can be viewed positively. The re-ordering of a room (such as a dining room, study etc.) into a bedroom – either within the spousal home or in the home of a non-spousal carer – can be seen as a positive affirmation of the strength of the affective ties that underpin the caring relationship (Exley and Allen 2007). However this change in function of a room can be difficult where space is limited or where the spouse has to move out of the marital bedroom to make way for equipment necessary to support the care of their partner. As one informal carer explained, 'we needed a hospital bed which was in the alcove [in the lounge]. So she had a hospital bed and everything was geared. So we made a hospital room if you like' (Exley and Allen 2007, 2321).

The differing requirements of home and work for older people, informal and formal caregivers mean that the physical and symbolic meaning of the home must constantly be negotiated as both a site of care and of social and personal life. The significance of home as a social space, for example, points to why healthcare providers may encounter resistance from older people and their families in their attempts to reorganise domestic settings to accommodate the healthcare needs of the care-recipient and to satisfy health and safety requirements (Phillipson 2007). Hence the desire to improvise or subvert the logics of care aids in order to retain a sense of home produces an ambiguity of place for both carer and care-recipient – one that brings home and care into tension as the aesthetics of health care systems jostle against the aesthetics of home.

Home, care and bodywork

Central to care in the home is its equation with socially intimate and affective relationships as a prerequisite for individually tailored care. Bodywork challenges this. As Exley and Allen (2007) point out, one of the fundamental difficulties with contemporary conceptualisations of home and care in western society is their failure to recognise that historically, not only did a whole range of bodily and biological processes become increasingly individualised and privatised, but community knowledge about bodily care was also increasingly appropriated by professional carers. Trained nursing and medical care gradually replaced local knowledge and performance of care and healing. In the process, bodily care became increasingly invisible. This has been compounded by the fact that care professionals tend not to highlight this aspect of their work, focusing rather on their technical expertise and skills. So while the provision of intimate bodily care from a familial adult to a child is still deemed as acceptable, it is no longer the norm in close adult relationships.

Yet as ageing in place shifts care and support back toward the home, informal carers are once again taking on many of the routinised tasks of caring for the ageing body. This includes much of the intimate and personal bodywork involved in care, such as washing and bathing, dressing, toileting and feeding the care-recipient. Helping an older relative to perform normally personal and private acts such as bathing and toileting, for example, gives rise to transgressions of contemporary social taboos around care in western society – particularly cross-sex care. Not only can such activities create substantial work for the informal carer, but performing these aspects of intimate care can involve considerable physical and emotional distress for both carer and care-recipient (Wiles 2003). Whilst the transgression of these social taboos may be less acute in spousal care-giving it can be particularly difficult where an adult child is providing personal care for a frail older parent of the opposite sex. Policies designed around ageing in place thus need to recognise that relationships associated with home can be altered and challenged by the process of caring. As Lawton (2000) and others point out, this can severely compromise intimate relationships and can be particularly difficult where the informal carer retains a distinct memory of how the care-recipient was before his or her bodily decline. For informal carers – particularly spousal carers – this can lead to an acute sense of loss well before transition to institutional care or death occurs.

Given the social taboos that now mark the social boundaries of bodywork in western society, the more detached stance of the professional carer may be important in helping to make it more manageable (Lawler 1991). Hence the management of the care-recipient's body – and who undertakes that management – is critical to the construction of home as a caring space. It is a body that is subject not only to management by informal carers, but one that has also been assessed by formal care services in relation to the quantity and nature of care it should receive against some institutionally defined norm that will shift and change with prevailing policy. Yet as Dyck et al. (2005) point out, it is important that this assessment of care is not interpreted solely in terms of meeting the medically defined needs of the corporeal body; if 'social death' (Lawton 1998) is to be avoided the home also needs to understood as a place where valued aspects of the social body can also be nurtured and preserved. In other words, the social and emotional needs of the ageing body also need to be recognised and met. This requires the giving of socially sensitive care in a way that not only enables the care-recipient to construct and maintain amenable personal relationships within negotiated boundaries, but which recognises that care within the home goes beyond just meeting the physical needs of the body. Percival (2002) for example, points to the way in which designers of specialist housing for older people often assume the ageing body is relatively static and hence requires only limited space. Indeed, one older person in his UK study noted: 'So far as they're [housing provider] concerned, you don't need a lot of space. You are disabled so you should sit in the chair and that's it. You won't need space for moving around' (Percival 2002, 740). Such a perception, he maintains, can impact on an older person's ability to socialise or invite people for a meal as such places are generally too small to accommodate a table and allow circulation space for those who may require the use of a zimmer frame or wheelchair.

Hence, it is only through recognising the home and body as interrelated sites and scales of analysis, that are both fluid and constantly in process, that we can gain real insight into the complex structuring of the relations that shape experiences of care (Dyck et al. 2005).

The home/care dichotomy

Whilst the (re)domestication of care serves a range of political and professional agendas, it is also based on an idealised version of care in the family. As such it runs the risk of over-romanticising notions of care that privilege the home (Rowles 1993; Exley and Allen 2007). Not only is there is a danger of placing too much emphasis on familiarity and emotional attachment to place as a component of residential preference, but it is important to recognise that not all older people are attached to their home or community in the way that policy assumes. The discourse of ageing in place, for example, is underpinned by assumptions that the home is a place that enables self-expression and identity – a social space where individuals can be relaxed and at ease; where privacy can be maintained and where choice and autonomy can be augmented – ultimately, a place where an individual's needs can be best met. It privileges the value of caring relationships without acknowledging the interaction of pre-existing social relationships with the actual work of caring. Thus, it is also based on the assumption that the home is synonymous with loving relationships – in other words it assumes that caring *about* is the foundation for caring *for*. The home is thus privileged over the institution as the preferred site of care, with the latter often being characterised as a place where both the needs of the institution itself, and of group living, put a strain on caring ideals.

We should not forget that for some, the meaning of home represents a site of fear, physical and mental abuse, neglect and/or violence (Warrington 2001; Meth 2003; Blunt and Varley 2004). Inevitably the private nature of the home – and by extension policies designed to facilitate ageing in place – is likely to impede the detection of abuse. Indeed, a recent review of studies looking at elder abuse across a range of mainly, but not exclusively, western countries found that around one in four vulnerable older people who are dependent on care and support from informal carers are at risk of psychological abuse, with a further 5-6 percent subject to physical abuse and neglect (Cooper et al. 2008, 158). Further, in a recent study of 370 older women in the UK, Bonomi et al. (2007, 35) found some 3.5 percent of these women had been physically or sexually abused by their own spouses. Yet it is important to note that abuse is not restricted to family carers, indeed Bonomi et al. went on to note that one in six professional carers commit psychological abuse with one in ten committing physical abuse. For a small but important minority of older people then, the promotion of policies designed to support ageing in place will not equate to 'good care'.

Idealised assumptions about the home/care relationship can also mask its inherent tensions (Exley and Allen 2007). It can bring with it contradictions and strains that are often unacknowledged and which create challenges not only for older people, but also those providing their care and support. As already suggested, tensions exist between home as habitat and a site of social and emotional expression, and home as a place within which the mechanics of care work, and the regulation attached to its performance (particularly formal care) are enacted. The home/workplace dichotomy is, perhaps, exemplified by recent smoking legislation in the UK that prohibits smoking in the workplace. Care providers are now in the position of having to request that care-recipients refrain from smoking in the private space of their own home during periods in which formal care work is taking place.[1] How enforceable this legislation is within the private space of the home has yet to be determined, but clearly any suggestion that non-compliance could result in service withdrawal will be of significant concern to both informal carers and care-recipients. It would however be exceedingly difficult to apply and monitor such legislation in relation to the performance of informal care – particularly co-resident care. Yet whilst formal care work is subject to health and safety regulation, no such regulation applies to the work performed by informal carers. So whilst ageing in place is contributing to an increased porosity of the boundaries between home and work in the field of care, it also highlights the inequities that exist in the conditions of formal and informal care work within the home. Any attempt to construct informal carers within a co-worker model of care is thus riddled with contradictions.

In addition to the tensions between home and work, it is also import to critically evaluate how older people, themselves, perceive the relationship between home and the place of care. As Rowles (1993) and others have pointed out, many older people assess the benefits of ageing in place in an entirely pragmatic way. That is, their key concerns are focused around issues of cost, comfort and convenience and as such have little to do with any sense of physical, social, or emotional attachment to place. Rather, for an older person, the ability to remain in the domestic home, located in a community that he or she is both knowledgeable about and integrated into, is meaningful because it enables them to call on local networks of care and support for practical assistance from friends and neighbours rather than having to rely on formal care options. It is also worth pointing out that while there is substantial evidence to suggest obligatory relocation can have adverse health consequences, (e.g. Ferraro 1983; Pruchno and Rose 2000; Keister 2006) the negative consequences of relocation will not be the same for all older people. Where home represents a site of loneliness, fear and abuse, the experience of relocation may well be manifest in improved well-being.

One final dichotomy around home and care focuses around the family and how to balance the right of the older person to age in place versus the cost to the informal carer. Whilst legislation in the UK now acknowledges the rights of carers, as Tinker (1999) points out, this is not just a legal matter. For example, whose rights and desires should prevail when an older person wants to remain at home but the informal carer is concerned about risk and danger, or the family feel that they cannot provide an adequate level of care? These are ethical dilemmas that, in the UK at least, have yet to be fully tackled by policy makers, though new forms of [technologised] care may go some way towards redressing this dilemma.

---

1   Whilst clearly it would be far better not to smoke at all, the reality is that (at least amongst current generations) many older people still do – and see little benefit in attempting to stop 'at their age'.

Concluding Comments

This chapter has been concerned with the [re]location of care to the home. Firstly, it has sought to contextualise the current emphasis on ageing in place within a shifting ideological and policy context around care for frail older people and informal carers in the UK. This set the scene for a more critical discussion about home as a site of care for older people. Whilst the meaning attached to home is viewed as facilitating older peoples' ability to successfully age in place, the delivery of care within the home also draws into tension the home/work dichotomy. The increasing porosity between formal and informal care work, between public and private space and importation of the clinical artefacts of care within the home all challenge the meaning of home. This then, raises questions about institutionalisation of the home and the extent to which ageing in place is shifting the meaning of home such that it comes to represent just another part of the extitution.

Chapter 6

The Impact of New Care Technologies
on Home and Care

New and emerging care technologies are one of the visible, material signs of attempts to solve a range of health related problems in advanced capitalist economies. Given the current and projected growth of those in the older age groups and policies designed to support 'ageing in place', many of these technologies are targeted at buttressing the care needs (or perceived needs) of older people requiring support within the home (Barlow et al. 2005). Governments across a range of advanced capitalist countries – including North America, Europe, New Zealand and Australia – have indicated that increased use of technology is a key plank in their strategies for addressing the care needs of the growing numbers of older people (Department of Health, UK 2008a; Hogenbirk et al. 2005; Ministry of Health, New Zealand 2008; Mort et al. 2008). Support for this strategy in the UK is evident in the Government's announcement of an investment of some £80 million through a two year ring-fenced Preventative Technology Grant (initially from April 2007-March 2009, then extended until 2010) designed to initiate changes in the design and delivery of housing, health and social care. While these care technologies have the potential to enhance and maintain the well-being and independence of a wide range of individuals who would otherwise be unable to live independently in the home, they are also seen as part of a strategy to reduce the number of older people entering residential care and hospitals (Bayer et al. 2007). Indeed, the UK Government specifically stated its belief that the Preventative Technology Grant could reduce these numbers by some 160,000 older people (Department of Health 2008a).

This 'technological fix' opens up some exciting possibilities for enhancing people's ability to age in place, but it also raises some important questions. How, for example, do older people experience these technologies, and how might they be reshaping the nature of care performed? Whose needs are being met and who benefits from the development and implementation of new care technologies? Importantly for this book, to what extent are these care technologies effecting a reshaping of the landscape of care? These questions underpin the discussion in this chapter. Organised around four main themes, it firstly considers how new care technologies are changing care interactions and how older people requiring care and support are experiencing these changes; secondly, in the drive to implement these technologies, the chapter considers whose needs are being met and hence, who the main beneficiaries are; thirdly, it considers how these technologies are impacting on the nature and form of care work; finally, it considers how new care technologies may be reshaping the spaces in which care takes place. Before discussing these themes, however, the chapter gives a brief overview of the sorts of care technologies to which it refers.

New Care Technologies – An Overview

Care technologies include a broad spectrum of care 'support' encompassing devices and systems that either enable individuals to perform tasks they would otherwise be unable to do, or increase the ease and safety with which these tasks can be performed (Cowan and Turner-Smith 1999). These technologies can be (and often are) used within domiciliary-based e-health – including telehomecare and smart homes. The latter encapsulate non-obtrusive disease prevention and monitoring within domestic space, whilst the former describes how technology can improve existing home-care services (Demiris et al. 2004). Tinker et al. (2004) further distinguish between what they refer to as portable assistive technologies, (such as alarms, monitors, motion detectors etc.) and fixed assistive technologies that may require housing adaptation (such as lifts, ramps etc.). Technologies designed specifically for domestic space are frequently referred to as domotics.

Whilst acknowledging that a wide spectrum of care technologies exists – including assistive devices such as hoists, canes and rails that been commonly available for many years, this chapter focuses on newer and emerging care technologies such as those being designed and developed by large international companies that specialise in the development of information and communication technologies (ICTs). These range from:

- environmental control solutions (such as wireless control for electronic equipment in the home including television and video equipment, hands-free telephones, lighting and door systems that can be linked to telecare solutions);
- remote telecare and diagnostic systems (for example, through internet and webcam technology);
- electronic pill dispensers (designed to dispense tablets at pre-set times and setting off an alert call if not taken);
- wearable or 'smart home' devices designed to monitor and gather continuous data – for example, motion detectors (to monitor inactivity, falls, frequent use of household appliances and facilities such as fridges, food cupboards, toilets etc.) and intruder alarms;
- 'smart clothing and fabrics' with inbuilt sensors that can monitor an individual's health status such as heart rate, abnormal heart activity, pulse, temperature etc.;
- electronic monitoring and tagging devices designed to identify patterns of movement and location; and
- robotic pets designed to address social isolation and to address some of the emotional needs of older people.

As evident from the above, companies are actively designing a whole range of hi-tech solutions to both monitor the physical health and activity patterns of (primarily) older people and to support their physical and emotional ability to age in place.

New developments in wireless networks and mobile phones are further facilitating the development of these new care technologies through the creation of a consumer platform for them. The growing familiarity of these consumer technologies make it easier for a new generation of older people and their families to adapt to them. The growth of wireless networks, in particular, not only offers the potential for replacing existing wired home technology, but they are also more flexible to install, less costly and cheaper to maintain (Tinker et al. 2004). Hence, they are an increasingly attractive proposition for care providers.

New Care Technologies and the Changing Nature of Care Interactions

While new care technologies have much to offer, we cannot accept their development and implementation uncritically. To begin to understand what impact the widespread adoption of these technologies is likely to have on the landscape of care, it is important to consider how they impact on the end user. There is little doubt that at their best, hi-tech solutions can offer older people a level of control and independence in their lives that they may not otherwise have enjoyed. Older users of new care technologies, for example, have pointed out that even being enabled to undertake simple tasks such as switching on a light, opening the door or closing the curtains without having to rely on a carer both increases their sense of independence and inclusion (Mort et al. 2008). Work by Essén (2008) suggests that some forms of non-intrusive monitoring can also increase older people's sense of safety and security in the home. New care technologies have the ability to monitor for falls, movement, eating patterns, irregular heart activity and so forth, to ensure that lone dwellers or older households in which both partners experience frailty, can maintain as healthy and independent a lifestyle as possible, enhancing their ability to remain in their own homes for longer. Proponents of new care technologies thus make significant claims about their ability to increase independence through decreasing reliance on formal and informal care (see for example, Gustke et al. 2000; Demeris et al. 2001; Hogenbirk et al. 2005; Essén 2008). Proponents have also argued that their ability to monitor the older person can act to reduce stress and anxiety amongst informal carers. Others, however, argue that they simply create new or different forms of dependence, as the knowledge that the individual is being checked and monitored – albeit remotely – means that older people using these technologies are simply shifting their dependence from physically present human care to remote care systems (Mort et al. 2008).

So while new care technologies may enhance an older persons' ability to 'age in place' this needs to be balanced against the cost of increased dependence on new forms of care; whether the benefits of this new dependence outweigh dependence on human caregivers; and whether this is a desirable outcome. Indeed, as the following excerpt from a group discussion with older users of new care technologies in the UK reveals, there are some aspects of care that technology can never replace:

> MN: Technology can never replace human contact and carers. Carers come to 'Freda' and I think apart from the practical things that they do for her, I hear laughter going on all the time in the social interaction. And the laughter – that's so important and cannot be replaced by technology.
>
> JF: Technology has never replaced a light bulb! There are things that just need doing on a regular basis.
>
> FN: They can't give you a bath can they?
>
> MN: *Yes, it has its part to play but it's not a substitute.*
>
> (Milligan et al., forthcoming, authors' emphasis)

Concern has also been voiced that these new forms of care could result in a decrease in social contact. Whilst informal and formal carers will still be required to aid personal care such as dressing, bathing and toileting, new care technologies enable remote diagnosis and remote monitoring – reducing the need for face-to-face care by nurses and clinicians. Remote monitoring of an older person's activity patterns though internet technology (e.g. the 'Just Checking' system – www. justchecking.co.uk etc.) can also alleviate informal care-givers' concerns about their older relative. Whilst this is clearly beneficial to the informal carer, it also opens up the potential for reducing the extent of face-to-face contact between an informal carer and care-recipients. Concerns have thus been voiced that whilst new care technologies may help to alleviate stress amongst informal carers, they may also contribute to an increase in social isolation and depression amongst older people receiving care and support at home. So on the one hand, these technologies can be seen as having a role to play in enhancing the ability of older people to manage their lives within their own homes, but on the other, they hold the potential to exacerbate exclusion and isolation.

This raises the issue of whether – and what forms – of care technologies are seen as acceptable by older people. For many, these technologies appear to fall largely into two groups – those designed to enhance an older person's ability to manage their own daily lives (facilitative new care technologies) and those designed to monitor health and activity (surveilling new care technologies). As this excerpt from a group discussion with older users of care technologies illustrates, sensors, webcams, smart clothing and so forth that monitor and gather data on health, eating, levels of activity and so forth, that is then remotely collated and analysed, can raise significant concern.

I: What about the sensors for when you get up and down and go in and out?

CW: I wouldn't like it if someone came in and put in this, that and the other – I would feel a bit like 'big brother' was watching – like it's an invasion of your territory.

HF: … there are ways that this technology could be quite 'big brother-ish'.

MN: At all costs that should be avoided. If we can have systems that are private to us should we choose to use them – this other point is very valid – I wouldn't like it if somebody came and put that in – I would feel like big brother was watching – like an invasion of your territory.

(Milligan et al., forthcoming)

This can be particularly disturbing for older people with dementia. This was highlighted by one individual working for major Dutch company producing new care technologies. Commenting on one pilot for dementia suffers he remarked:

> After six months, they switched almost everything off, because a lot of people were very afraid of a voice coming out of the wall, and a camera that's continually following you … So the people were continuously very disturbed by all the technology that they saw (Mort et al. 2008, p.68).

Resistance and subversion

Older people requiring care and support can often feel a clear lack of control over their own lives – a feeling that can be exacerbated by surveilling technologies and reliance on remote care technologies over personal attendance. Interestingly a number of studies have revealed how some older people seek to demonstrate agency and regain control by actively resisting these kinds of technologies. As participants in a study by Milligan et al. (forthcoming) noted:

JF: Just before I retired, I was a nurse, and tablet dispensers were the ones that bothered me, because I know that when I was going in to elderly patients' homes to remind them to take medication that there was a lot of joking around this, and I had to cajole them into taking it, 'now come on, you'll get me the sack if you don't take it' and all those kind of interactive things that no machine could possibly do.

I: Some papers describe how people open and shut them to make it seem that they have taken them.

KW: That in itself is an assertion of your independence – to say I'm going to shut it.

Research has also drawn attention to the way in which older people purposefully seek to 'subvert the system' through varying their routines or adapting the use of the technology to see what will happen (see for example, Tinker et al. 2003; Lundell and Morris 2004; Wu and Miller 2005). Wu and Miller (2005) note, for example, how some older people resist or 'mis-use' care technologies by developing alternate uses for them – such as playing games or using telecare systems in inventive and even disruptive ways. Significantly, monitoring technologies are often 'mis-used' to trigger the very social responses they are designed to reduce. While the adaptation of the technology often continues to provide the cognitive or monitoring activities the technology was actually designed for (albeit in different ways), such actions demonstrate how older people have found ways to adapt the technologies offered to them in ways that better meet their needs (Mort et al. 2008). This not only highlights the importance of understanding the environment within which new care technologies are to be located, but also that there is a clear requirement to ensure that the social and affective needs of older people do not become subsumed by their medical needs. Indeed, researchers have pointed to the importance of ensuring developments in new care technologies take seriously older people's ongoing and ever-changing need for meaningful human interactions. Work by Morris et al. (2003) for example, illustrated how older people with varying states of cognitive decline feel very strongly about loneliness and the need to maintain social ties. Meeting these social needs, they argue, is central to older people's health status. Interestingly, some care technology designers have begun to take the importance of addressing these issues on board (Morris et al. 2004; Pols and Moser 2009). These technologies help older people to monitor and broaden their social interactions, or express affection – for example, through stroking or interacting with a robotic pet. Significant claims are being made for the beneficial therapeutic effect of these new forms of care technologies (see http://news.bbc.co.uk/1/hi/technology/6202765.stm). Indeed, Pols and Moser (2009) go so far as to suggest that such developments have the potential to blur the divide between what has traditionally been seem as 'warm' (human-centred) care and 'cold' (non-human centred) care technologies.

Surveillance, privacy and new care technologies

As indicated, many new care technologies involve understanding older people's everyday routines and use alarms to monitor any deviation from this routine. Information about these daily routines is collected, aggregated and stored in databases. This raises important ethical and legal questions about consent and data ownership. For example, these data could be of great value to those involved in the design and marketing of new care technologies. Hence, collecting, interpreting and acting upon such data raises critical design issues linked to reliability, responsibility and informed consent (see Blythe et al. 2005). Important questions thus need to be asked about who has access to this remotely gathered data and the ethical implications of this (for example, around privacy and consent). But privacy is more than a concern about third-party access to electronic medical information; as Miller (2001) points out, it also includes psychological privacy (the revealing of intimate attitudes, beliefs and feelings), social privacy (the ability to control social contacts) and physical privacy (the ability to control physical accessibility). The nature of the two-way, interactive video telecare interaction, he suggests, is very different from the conventional, face-to-face encounter. Unlike face-to-face encounters, for example, video-consultations often involve more than one health professional; hence patients may feel more inhibited by the clinical encounter and more susceptible to privacy violations. Moreover, unlike face-to-face consultations, video-consultations can easily be recorded, and while this may allow for a more complete medical record, as Miller points out, it also increases the potential risks associated with third-party access, resulting in patients may be more anxious about their privacy.

This is particularly important in relation to communication and the exchange of information about the care-recipient among health and social care providers and informal carers. As Tracy et al.'s (2004) work on care-giving to people with early stage dementia in Canada noted, professionals valued disclosure of information to both colleagues and informal carers (justified as being 'in the patients' best interests'). Those with early-stage dementia, however, whilst accepting the need for disclosure amongst healthcare professionals, wanted the ability to exert strong control over disclosure to informal carers and other family members. Unsurprisingly, informal carers valued the opportunity to be kept informed of the care-recipient's condition, even if this came without the latter's consent. Yet as Morris's (2005) work with informal carers who had access to graphic representations of their ageing parents' behaviour patterns and social interactions revealed, this is not without its problems. Such access can impact not only on informal carers' relationships with their parents (who may feel aggrieved at having their privacy invaded by their own adult child) but also with the informal carer's siblings, where a sense of 'rivalry' about who is or is not participating in the care of the elderly parent may emerge. New care technologies, then, hold the potential to intervene in relationships previously thought to be private. Magnusson and Hanson (2003) maintain that this breaching of privacy represents a potentially 'unjustified paternalism' – one that needs to be treated with caution. Hence, new care technologies not only raise important questions about possible unequal or problematic caring roles but they also raise ethical questions about everyday interactions, privacy and consent between carer and care-recipient.

Critics, however, maintain that such interpretations of the impact of new care technologies have been overly concerned with threats to privacy and liberties. Indeed, they go so far as to argue that this is as a one-sided dystopic view that is both superficial and analytically unfounded (Lianos 2003; Blythe 2005; Essén 2008). Essén (2008), for example, points out that surveillance and control are integral parts of care and as such, they are both conceptually and empirically difficult to separate out. Furthermore, she maintains that the issue of whether or not surveillance and monitoring should be viewed as 'bad' is contingent on both the user-context and the agency of the surveilled subject. In direct contradiction to the comments raised by older users of new care technologies in this chapter, Blythe et al. (2005) assert that care technologies that monitor daily routines and habits have the potential to not only track changes in behaviour patterns over a period of time (e.g. changes in sleeping, eating or other activity patterns) but may be more acceptable to an older person as they are less obtrusive than, for example, other care technologies such as the wearing of a falls detector. Lyon (2006 2007) further argues that such technologies can, in fact, be enabling in that they are *less* intrusive than the alternative option of residential care. These are of course, important points – ones that should not be overlooked in the debate about the widening use of new care technologies in the care of older people. Yet in their attempt to redress the balance such critiques run the risk of over-compensation through minimising or over-simplifying complex considerations.

Disentangling care, monitoring and surveillance – whether at the human or technological interface – is of course extremely difficult and it is hard to imagine how we can give care without watching over those we care for. But we should not fall into the trap of assuming that the human act of watching over those we care for is always a benign process. Shifting power relationships are an inevitable part of the act of care-giving (Milligan 2003). This is not to infer that the power relationship is – or indeed should be – a one-way flow, yet as Twigg (2001), Milligan (2003) and others point out, as an older person becomes increasingly frail and reliant on human care, it can become increasingly difficult for that person to exert power, and therefore agency. Nevertheless, any discussion about the development and implementation of new care technologies needs to engage with the views of a wide range of end-users of these technologies – particularly in relation to what they consider to be empowering and disempowering technologies. Further, these discussions should not be set against fears that residential care is the only alternative option, rather they need to be framed within debate about what constitutes good care for older people and how new care technology can contribute toward the construction of a more enabling society for both older people and their informal carers. That is to say that we should not think of care technologies as being somehow either all 'good' or all 'bad', rather we need to embrace their potential whilst at the same time recognising their limitations and the impact they will have on care interactions.

For older people, new care technologies that are designed to enable them to make personal choices, to undertake more of the everyday tasks of daily living or which enhance their sense of security within their own homes are seen as far more acceptable than those which remotely monitor their health and activity patterns. But as Essén (2008) rightly points out, this needs to be interpreted within the individual user-context. Nevertheless, it seems clear that new care technologies need to be considered as an *aid* – not a solution – to any growth in the demand for care. This becomes increasingly relevant when we also take into account the additional – and perhaps less overtly recognised – role that both formal and informal carers play in the home. That is, that the physical presence of a carer within the domestic setting enables them to monitor the cleanliness of the house, the need for repairs, the state of the garden, the amount of food in cupboards and fridges etc. – aspects of the older person's ability to deal with the general upkeep of the home and the management of their everyday lives that cannot be picked up by remote monitoring. As one older user of these technologies in the UK put it, 'what about "one picture's worth a thousand words"? If they just look at you they can tell very often, far better than talking to you' [on the telephone] (Roberts and Mort 2009).

Access

Whilst there is a growing familiarity with the sorts of platforms that support at least some of these new care technologies, there are significant disparities in their availability, accessibility and affordability within and across different countries. Cabrera and Rodriguez (2005) for example, point to the greater prevalence of new care technology usage in Scandinavian countries in comparison to Southern Europe. Others point out that where healthcare systems are private or insurance based, access to these technologies will largely be limited to the more affluent members of those societies. Given that older people tend to have substantially lower incomes than younger age cohorts within similar socio-economic groups, access is likely to be disproportionately biased. Where state welfare systems exist, access to new care technologies is likely to be more evenly distributed, but the forms of technologies on offer may be limited. The implementation of new care technologies following costly, but successful, 'all singing all dancing' pilots programmes, for example, are often based on cheaper, stripped back version of the system piloted (Mort et al. 2008).

Accessibility is more than a question of differential access across geographical space. For current cohorts of older people, at least, it also revolves around the challenges of dealing with a technology that has become available relatively late in life. Wu and Miller (2005), for example, point out that older people can face particular difficulties with interactive devices such as touch screens, computer and telephone interfaces that are typically installed in smart home technology as a consequence of both physical and psychological barriers. Chronic conditions of older age, such as dementias, stroke and arthritis also present particular challenges (see Alm et al. 2004). As Torrington (2009) points out, this requires careful design that involves the early engagement of older users if they are to be both accessible and successful in terms of meeting the needs of older people located in different places. Despite this, a number of studies have indicated that where the technology *is* available and seen as useful to their everyday lives, older people are able to successfully adapt to it with appropriate training (Demiris et al. 2004; Magnusson et al. 2004).

New Technologies and the Changing Nature of Care and Care Work

The implementation of new care technologies has implications not just for older people requiring care and support, but also for those involved in the delivery of that care. Many of these technologies, particularly tele-clinics and call-centres, not only change the patterns of care, but they bring new sets of actors with new sets of skills into the care network. Nursing and medical staff using webcam technology, for example, are required to learn new diagnostic skills that operate in isolation from any visual, tactile or olefactory cues they may normally employ. Call centre operators require listening and decision-making skills that are removed from any visual cues a face-to-face care worker can normally employ to support their care work. Telecare installers and technicians require social and explanatory skills in order to assess the needs and difficulties an older person may have – either in understanding how these care technologies work or in assessing why a technology may be failing and what might be required to meet the changing care needs of the older person. Informal carers may also be required to learn to use and interpret remote web-based and mobile technologies.

Commentators thus point to a shift in work activities that operates in a downward cascade from hospital-based clinicians to general practitioners (Brown 2003) from doctors to nurses (Merresman et al. 2003; Martin and Coyle 2006; Engstrom et al. 2005); from nurses to call centre staff (Soopramanien et al. 2005); from clinicians and nurses to patients and informal care-givers (Oudshoorn 2006). These changing patterns of care work bring with them new care responsibilities: nursing and other clinical staff, for example, may have to make remote medical assessments via videoconferencing (Laflamme et al. 2005; Mahoney et al. 2001); remotely located call centre staff are now charged with making decisions about the appropriate response to automated calls (Lopez and Domenech 2006); technologists responsible for the installation and maintenance of new care technologies can also be charged with making decisions about what technologies should be installed (pers. comm. 2008); informal carers are charged with monitoring their older relatives behaviours patterns and respond to automated calls (Tracy et al. 2004); older people are required to make decisions about when to take medical measurements in the home with the data being remotely transferred to distant care settings (Oudshoorn 2006).

These changes act to release some groups from some aspects of care work, whilst allocating new activities to others. This redistribution of care work may or may not be burdensome, but it does have implications for the gendered nature of care. As the division of labour in the health field is still notably gendered (with women typically clustered at the lower ends of labour hierarchies), these cascades have the potential to further reinforce the gendered nature of care. As Chapter 3 has already demonstrated, care for older adults is still primarily a female responsibility; gender inequality and the gendered nature of the labour market in employment policies is likely to compound this situation (Rubery et al. 1999; Jönsson 2003). Hence, new care technologies are likely to both depend on, and reinforce, existing gender relations within the reconfigured landscape of care.

A New Technological Topology of Care?

Policies focused around 'ageing in place', together with the growing sophistication and implementation of new care technologies as part of – or even a replacement for – existing packages of support for older people living at home require us to think about how these technologies may be contributing to a shifting landscape of care. Here, two strands of thought are of relevance: firstly, the extent to which these technologies may be contributing not only to a reshaping of the domestic home, but also how older people identify with the reconfigured home; and secondly, the extent to which new care technologies may be creating a spatial re-ordering of care work and care practices.

New care technologies and the reshaping of the home

As Blunt (2005) and others have noted, the home is both a material and affective space – the meaning and experience of which is both shaped and reshaped by everyday practices, social relations, memories and emotions. At its best, it is seen to offer security, familiarity and nurture (Tuan 2004). The drive toward ageing in place is thus underpinned not only by perceptions that the home is where older people are likely to feel most independent and in control, but also on the premise that the home contributes to a person's sense of security and identity. In acknowledging that the home often (but not always) represents a form of sanctuary for older people requiring care, Friedewald and Da Costa (2003, 28) point out that in integrating new care technologies into the home it is important that the technologies do not dominate the overall function and experience of the home. Rather they should seek to 'enhance the quality of life of residents, not only by facilitating their daily activities, but also supporting their socialisation'. In other words, the physical manifestation of new care technology devices should aim to be as unobtrusive as possible and be designed to meet both the social and medical needs of the care-recipient. This is important, as researchers have drawn attention to the ways in which the implementation of care policies and practices designed to support ageing in place can also create changes in the meaning of home and how people identify with home (Milligan 2000 2003; Twigg 2000).

Yet older people can also view what might be considered 'everyday' technologies such as televisions and computers as intruding on the way in which they identify with home. Dickinson et al. (2003) for example, point to instances in which older people have sought to cover televisions and computers with cloths when not in use in an attempt to reconstruct the physical appearance of these technologies in a way that blends with their perception of home. This may also infer that, for these older people, technology is not part of their everyday life. One respondent in Dickinson et al.'s study, for example, noted her difficulty in 'getting down to the controls' on the video recorder but was dismissive of the remote control, whilst another had arranged a number of remote controls in a 'hub' around her armchair. Photographs and memorabilia of key events tended to be displayed in areas of the house where visitors would circulate. Many participants were prescribed daily medication and struggled to remember to take it. Some developed coping strategies such as leaving the medication in a single location, placing calendars, notes and prompts in strategic locations. Such observations suggest that older people's requirements are not just surface needs linked to safety and security, remembering to take medication or to eat lunch etc., rather such needs, and how they are *felt*, are embedded in a lifelong sense of individual and social identity.

Earlier sections of this chapter have illustrated how new care technologies that facilitate an older person's ability to manage everyday tasks are seen as welcome and enabling devices. Environmental control systems based around a central hub and which enable a person's ability to answer the door, open and close curtains, switch on lights, or entertainment systems etc. are both relatively unobtrusive and enabling to older people. Monitoring and surveillance systems are viewed as enabling to informal and formal care-givers, and whilst relatively unobtrusive in a physical sense (consisting largely of motion monitors and so forth) they are clearly deemed to be more affectively intrusive to those requiring care and support. For some this has resulted in behavioural change within the home as older people seek to 'test' or challenge the system. As also suggested above, responses to the intrusiveness or otherwise of monitoring technologies is highly individualised and contextually dependent.

For older people and their informal carers, the affective experience of home can be as important as the physical structure (Milligan 2005). Enabling older people to respond to the use of new care technologies in the home thus requires policy makers to recognise that design needs to take into account the ways that technologies may shape the physical and affective aspects of the home. Heywood (2004) cites a range of literature which points to the detrimental impact upon health when professionals involved in the delivery of adaptations fail to consider psychological factors and the meaning of home to recipients. As Heywood (2004) notes, when unwelcome adaptations are installed in the home, recipients can feel helpless and disempowered. How new care technologies act to reshape the home and people's experiences of being 'at home' is thus critical to development of 'good care'. Friedewald and Da Costa (2003, 18) further caution against the over-usage of new care technologies within the home, maintaining that, 'If initiative or physical movement is no longer needed, the "passivating" implications for the elderly, especially, might prove deteriorating to one's physical and mental health.' The challenge, they argue is to produce high quality care technologies that also provide for safety, stimulation and socialisation. Critiques of new care technologies rightly point out that the home is more that an array of technological tools – the function of which is to help older people requiring care and support to survive in their daily lives, rather, 'Home is for humans, whose quality of life is expected to improve via technology and ambient intelligence. Home is an emotionally charged and personally furnished cradle of living – physical space as much as a socio-cultural context and a state of mind' (Friedewald and Da Costa 2003, 19).

Arguably, the concept of 'home' is becoming more broadly defined to include appropriate access to services and neighbourhood facilities as well as good quality secure accommodation. Barlow et al. (2005) suggest that this is beginning to challenge notions of what type of home is appropriate for people's varying needs. Given that telecare is able to facilitate the delivery of care to widely dispersed properties, the focus of care no longer has to be on single sites. At its best, then, telecare has the potential to facilitate the delivery of more user-focused care to people in their preferred environment.

It is of course important to reiterate that home does not represent 'safety' and security for all older people, particularly in the light of elder abuse (Taylor et al. 2006; McGarry and Simpson 2008) nevertheless it is important to consider the extent to which care technologies can change the experience of home, not only for older people but for societies as a whole. The challenge, then, is to promote cost effective, user friendly technology that responds to users' needs and remains sensitive to the socio-cultural context of 'home'.

The spatial re-ordering of care practices

Given that new care technologies are socio-technical innovations designed to address the care needs of older people within their own people's homes, these technologies open up the potential to reorder care practices and relationships. In part, they represent technologies of deinstitutionalisation in that they both enhance older people's ability to age in place, and facilitate clinical and care encounters away from traditional institutional settings. These encounters can occur in more diffuse and disparate spaces including not just the domestic home or the clinical setting, but also the call centre and even the internet café. Indeed, the advent of wi-fi technology holds the potential for web-based monitoring of older people requiring care to occur in almost any remote setting where a wi-fi connection can be made. Commentators such as Domènech and Tirado (1997), Domènech et al. (2006) and Lopez (2006) have argued however, that community health care and call-centre based home care for older people are indicative of the *extitution* in that they seek to control (rather than actively discipline) patients and users through processes and programmes rather than buildings or enclosures (Mort et al. 2008).

Hence, in thinking about new care technologies, we have to recognise that while much of their physical manifestation is, indeed, writ within the home, they also brings into play *new* sites of care that can be remote from both the home and the institution. Call centre, teleconference, telediagnosis and monitoring stations, for example, all involve sites of care that are linked to, but remote from, both the home and traditional institutional arrangements. In other words, they are illustrative of extitutional arrangements in that they comprise both the virtual and material arrangements of care and thus illustrate a clear end to the traditional interior/exterior distinction identified by Vitores (2002). In other words there is no one building that these new care technology services inhabit, rather they are part of the network of care that is located in a series of places that are both dispersed across material and virtual space. Whilst remote monitoring, tele-diagnosis and webcam technology, for example, facilitate the delivery of care within the domestic space of the home, it does so from spaces that are distant and remote from the care-recipient.

Concluding Comments

This chapter has sought to draw attention to the ways in which increased emphasis on new care technologies is beginning to change the landscape of care away from traditional institutional and community-based arrangements to extitutional ones, in which new care providers in places remote from traditional care settings are drawn into the care network. This, it is suggested, has implications for who cares, the form of that care and the changing spaces in which it occurs. Whilst I would maintain that the notion of the institutionalisation of the home still holds validity – particularly for the most frail elderly – this has to be set within these new arrangements for care that take us beyond both the home and the institution. Within the home we are seeing an increased porosity of the boundaries between public and private space and between home and work.

Finally, whilst it is clear that older people may welcome enabling technologies, they can also hold significant reservations about those that can be seen as monitoring or surveillance technologies. Informal and formal carers on the other hand may see significant benefits in the use of monitoring and surveillance technologies. This highlights the tensions that can exist between the needs of the care-recipient and those of the carer – raising questions about who these technologies are, in reality, designed to benefit. Critically, new care technologies need to be seen as an adjunct to, not a substitute for, human-based care.

Chapter 7

Care and Community?

The domestic home does not exist in a vacuum but is always connected to other places. Indeed, Moss (1997) maintained that the home represents both a fluid and shifting space – one that extends beyond the physical boundaries of the building itself to include community sites, spaces and amenities as well as friends, neighbours, those working within the community and so forth. Its relative location is, thus, important because it directly affects access to other locally-based people and resources that can support both informal carers and care-recipients. Hence this chapter draws on Moss's extended notion of home to examine issues of community, care and support outside of the domestic (familial) home. In part, this refers to experiences of, and access to, such services as day hospitals and day care, lunch and activity clubs and so forth, that are provided by a range of statutory, private, or voluntary care providers. In part, it refers to the community itself. Community, here, is understood as both as an enabling geographical neighbourhood and as the social actors that makes up that neighbourhood. The underlying ethos of community care policy, for example, assumes that the local community within which care takes place will also act as a supportive 'prop' to its frail and ageing citizenry. Neo-liberalising policies, in particular, have sought to address this through attempts to reinvigorate active citizenship within local communities. Such policies, then, presuppose that frail older people and their informal carers will be able to draw on local social networks as a means of supplementing care and support.

How these places are experienced by frail older people and their informal carers is thus important if we are to understand what facets of community-based support facilitates or constrain people's ability to age in place successfully. Yet it is also essential to recognise that the familial home is not the only form of residence for older people residing in the community. Different forms of supported accommodation ranging from traditional Sheltered Housing to the newer 'home for life' retirement village are indicative of the growing range of housing alternatives designed to support ageing in place. The chapter thus discusses the implications of new housing alternatives for the care and support of older people before finally taking a critical approach to the issue of community itself and what we mean by community in the twenty-first century. As suggested above, community care policy and ageing in place is based on the assumption that 'the community' is manifest through a fairly stable geographical locality around which services can be planned and within which older people and their carers can benefit from informal help and support. There is a growing literature, however, that suggests that the essential basis of this assumption is flawed. The social and economic processes of globalisation, these literatures maintain, may be impacting on the relationship between older people and the nature of care, community and place.

Community Spaces of Care?

Research has demonstrated that with increasing age and frailty, the spaces that older people inhabit decline and tend to be increasingly focused around the home (Rowles 1987; Day 2008). Whilst wider spaces of social inclusion such as parks, shops, restaurants, cinemas, theatres etc. are of course important, frail older people are likely to be particularly sensitive to their immediate physical surroundings. Any environmental inequity in terms of accessibility to that environment, proximity of services and amenities is thus likely to impact negatively on the ability of these older people – and by extension their informal carers – to function in that environment. Indeed, Day (2008) goes so far as to suggest that that the physical neighbourhood holds the potential to affect the collective well-being of older people, pointing out that, 'surface materials, seat availability and design, steps and access routes, lighting, street layout, and signage have all been noted as potentially posing problems for older people' (301).

In large part these issues are linked to social and physical access and mobility. Day (2008), Wiles (2005) and others have pointed to the way that even relatively simple things, such as the availability and positioning of key services, the presence or absence of accessible toilet facilities and cafes in local settings and so forth, can severely curtail the geographical reach of both the informal carer and care-recipient. For informal carers, this can be exacerbated by the sheer effort involved in assisting a frail older person (who may have limited mobility and flexibility) to prepare for the outing and – if the destination is outwith walking distance – the need to arrange suitable transport to that site.

Much of the focus of community-based support for older people, however, has been concerned with spaces of care, respite and support. These include medical and day-centres, day hospitals and support groups; targeted social spaces specific to older people, such as lunch and activity clubs; as well as social spaces of personal meaning such as the homes of family, friends and neighbours. By extension they also serve as a milieu or background context for informal and formal carers and others living and working there. But how, as well as where, community services are provided is also critical. As Arksey and Glendinning (2005) have demonstrated, inflexible services, unreliable and expensive transport arrangements, inadequate levels of day care, and the lack of affordable and suitable after-school and holiday care for children within the local environment make it difficult for informal carers to combine paid work with informal care-giving. This means that informal carers often find themselves forced to transfer to part-time hours or to not work at all.

Gibson et al.'s (2007) study of informal carers' and care-recipients' experiences of community-based memory clinics in the UK has also drawn attention to the importance of unpacking the 'how and where' questions in attempting to understand the acceptability or unacceptability of services to both service users and carers. This, they maintain, is particularly true in the context of dementia, where anxiety and confusion can be integral parts of a disease process that is profoundly influenced by social and environmental processes. Drawing on their own study they further highlight how informal carers felt that whilst the relocation of memory clinics from institutional settings to the community was a welcome development, by and large such clinics simply acted to replicate the medicalised requirements of the old institutional clinic system, rather than operating to meet their specific needs. In comparison to memory services received within the home, travelling to the clinic was also viewed as problematic, particularly where people had to rely on public transport or ambulances. Such experiences were seen to heighten the perception that informal carers and care-recipients had to operate within the confines of a system that was not designed to meet their specific physical and psychological needs. In comparison, receiving treatment at home was seen to provide the individual with a greater perception of control and empowerment over their treatment – one that transcended the experience of community-based health care services.

As noted in Chapter 5, however, there is significant geographical variation in both the type and quality of services that are available – a factor that is manifest not just across local authority jurisdictions, but also between them (Milligan 2001; CSCI 2009). Work on community-based services in rural areas, in particular, has drawn attention to the challenges associated with community-based care provision in these settings. Researchers have, for example, drawn attention to the difficulties of recruiting care workers and hence the lack of choice in relation to who is employed to undertake that care work. Issues of isolation; travel and distance between care-recipients; limited choice and access in relation to services (particularly community-base services); poor housing conditions; and the increased challenges of delivering services in poor weather conditions have also been raised (Milligan 2001; Wenger 2001; McCann 2005; Skinner et al. 2009). Indeed, McCann (2005) goes so far as to suggest that in rural areas the effectiveness of rural community care is largely reliant upon the goodwill of the community itself. Such perceptions have lead Wenger (2001) to argue that rural community-based service provision demands a substantially different approach from that appropriate to urban areas. Indeed, she suggests that in such locations a community-based approach may be more appropriate. Even where specific forms of service provision are geographically proximate, limited mobility, poorly designed infrastructure or lack of adequate transport can inhibit an older person's ability to access those resources. Hence, the extent to which an older person is able to successfully age in place is an outcome of the dynamic between the competencies of the individual, the personal and social support available within that community, and the demands of the specific environment.

Cutchin's (2003) work on day-centres in the North American context is useful here in that it highlights the importance of scale in understanding older people's relationship with community-based care settings. Focusing specifically on day care, he maintains that community-based settings not only change the experiential context of place, but how day care is delivered will be facilitated or constrained by both the location and design of that setting. Busy suburban streets, he suggests, are more likely to facilitate integration into community life than so-called 'therapeutic rural settings' where care-recipients may feel isolated and transport links difficult. Focusing on the micro-scale, he further points out that many day-centres are held in buildings designed for multiple purposes – such as village halls, community centres, residential care homes, hospitals and so forth. Not only does this mean that there will be competing demands on their usage, but they may not be well suited to the needs of day care. Further, where day care is provided in care homes and hospitals it can create tensions. One the one hand, older residents may feel those attending on a non-resident basis are 'invading their space'; on the other, non-residents may feel uncomfortable being reminded of what the future potentially holds for them. Cutchin's work also points to the way in which some care-recipients can create territorial conflict through the 'colonisation' of space – for example, through claiming particular areas of the day-centre as 'theirs'.

But what exactly do we meant by community in relation to care? Community care policy in the UK is underpinned by the concept of a re-engaged citizenry that will work together for the betterment of their local communities (Fyfe and Milligan 2003). By-and-large this refers to defined geographic communities.[1] With regard to the care of older people this assumes that: a) it is better to be engaged than disengaged; and b) that interaction promotes well-being. Indeed, Cutchin (2003) maintains that the processes of interaction within community-based day-centres and so forth are designed to construct new communities of older people – to help them make new friends and experience a sense of belonging through the development of new social and activity networks. Meaning, he suggests, is produced in a reciprocal process of shared activity and interaction. In other words, older people attending day care help each other and in doing so, create bonds and meaningful relationships.

It is important, however, to recognise that older people are not a homogenous group so while these assumptions may well hold true for some older people they are not equally applicable to all. Day care, lunch and activity clubs and so forth can be important in terms of 'providing a change of scenery' for older people who would otherwise be housebound and offer a period of respite for the informal carer. Yet as Martin et al. (2005) point out, community-based facilities such as day hospitals and day-centres can also be characterised as alien and alienating environments of clinical domination and personal uncertainty. Further, they can also represent sites of enforced engagement. That is, that rather than a process of reciprocal interaction, the care-recipient can find themselves having to engage with

---

1   Although it is worth noting that UK social policy now recognises the benefits of tackling enduring some social problems thematically rather than geographically, by targeting communities of peoples such as Black and minority ethnic groups, teenage pregnancy, youth crime etc.

other older people that under different circumstances they would have chosen not to, in places they would not otherwise have visited, and participating in activities they would otherwise not have undertaken.

Not only do such community care services consist of a 'melting pot' of older people who may have little in common, but activities are often designed and organised by activity co-ordinators with little or no input from the older people themselves (Kontos 1998). This, in itself, can leave older people feeling actively disengaged and disinterested in the process. As one older care-recipient in a study by Bamford and Bruce (2000, 555) commented in relation to a day-centre: 'there's nowt wrong about the place don't get me wrong, but ... You get a bit bored.' At its most extreme, resistance to integration is manifest though an inability to tolerate difference. Cutchin (2003), for example, illustrates how some care-recipients can be demonstrably intolerant of those experiencing different (notably mental) conditions or noticeably poorer physical health – particularly those deemed to be more vulnerable or disruptive than themselves; or from different gender or cultural backgrounds. Hence such sites can reinforce exclusionary experiences for some frail older people. As Martin et al. (2005) are lead to comment, community care policy imperatives can manifest in complex relationships of power that exist within and between care-recipients, care professionals and informal carers in relation to specific care settings.

Care-recipients and their informal carers can also have different perspectives about the importance of integration into the local community (Bamford and Bruce 2000). This highlights the tensions that can exist between the needs of the frail older person and the needs of their informal carers. For care-recipients community integration is often about friendship and social networks. For informal carers, it can be about the provision of critical opportunities for them to gain a much needed break from caring. Community integration for the informal carer may also be more about the role of neighbours in monitoring the care-recipient, for example for evidence of activity, falls, or wandering in the case of dementia. The latter is underpinned by the notion that the community (as engaged citizens) cares *about* and thus will care *for* its older residents.

The notion of community care and community integration draws on romanticised images of a supposed community spirit that existed in the mid-twentieth century and that we have somehow 'lost' in recent years. There has, however, been considerable criticism of the supposed existence of such 'caring communities'. Indeed, Townsend's (1957) seminal work on older people in the mid-1950s noted that, in reality, people within most communities were very restrained in their relationships. Older people, he maintained, had few friendships outside the family. Recent work by McKibbin (1998) and Savage et al. (2005) on the general pattern of neighbourhood ties in working class communities in the UK during this period also indicates that nostalgic perceptions (particularly amongst older people) that some kind of enduring community and neighbourhood spirit existed are highly romanticised. As with contemporary society, unless people had known each other for many years, relations with neighbours tended to be superficial.

Care and Transition – From Home to Supported Accommodation

By and large ageing in place is equated with 'staying put' in the domestic home (or that of a close relative). Increasingly, however we have seen the emergence of a range of alternative housing options designed to offer differing forms and levels of care and support as physical and/or mental frailty increases. Whilst the move to such accommodation represents a transition in the life course of the older person, the conceptual thinking behind such developments is one of maximising 'independence' by delaying or eliminating the need for institutional care where the family home is no longer a feasible or desirable option. Though the form of alternative housing options and the terminology used may vary across different national settings, they represent part of the spectrum of 'places to live' with care and support that lie somewhere between the familial home and residential care. In the UK, these options range from Sheltered Housing and Extra Care (sometimes referred to as assisted living), to Retirement Villages and 'close care'. These options are offered by a range of providers including the private sector, charitable trusts and housing associations as well as partnerships between local authorities and non-profit providers.

Sheltered Housing for older people has been evident in various forms in the UK since the mid-twentieth century and generally comprises groups of self-contained flats or bungalows that can be offered either for rent or sale. As a general rule, Sheltered Housing also has communal space such as sitting rooms, laundries, garden areas and guest accommodation. Individual homes are usually fitted with some form of telecare such as entry and alarm call systems and may also have a range of other low level adaptive technologies. The complex of dwellings may also have an on-site warden whose role is to provide support, advice and guidance. Extra Care housing offers more intensive levels of (24 hour) on-site support and wheelchair accessible housing. In addition to those communal facilities found in Sheltered Housing, Extra Care can include a range of other amenities such as a restaurant or dining room, health and fitness facilities, activities and domestic support usually provided by on-site staff. At a conceptual level, Extra Care is seen to comprise a form of independent dwelling designed to be 'homelike', hence they should not look or feel in any way institutional (Evans and Vallelly 2008). Extra Care has been a key plank of recent policy around the care and support of older people with the UK government investing some £147 million between 2004 and 2008 in order to promote its development. A growing number of these dwellings are located in retirement villages that can consist of anything from an estate to a full-blown village-sized development with a range of self-contained bungalows, apartments and so forth. These complexes are designed specifically for older people and often located in relatively rural areas, though close to transport links. They are generally centred around a large central clubhouse (often a restored and converted mansion or stately home) that acts as the hub of social activity. As Croucher (2006, 1) put it, 'Retirement villages offer high levels of care and support in living environments that maintain and promote independence, with the additional benefits of a range of social and leisure activities.' These villages often include a residential care home that enables them to offer 'close care' schemes whereby the care home which will deliver personal care to a resident living in self-contained accommodation proximate to the care home. Retirement villages and 'close care' thus claim to offer the potential to facilitate a transition to residential care, if required, at some later date in a way that reduces the trauma and anxiety that can occur during transition from the familial to the residential care home. The claim of retirement villages is that they offer a 'home for life'– a safe and supportive environment in which an older person can live out the remainder of his or her days amongst a community of peers.

Proponents of these alternative sites of care maintain that whilst informal carers and other family members continue to play a significant role in the practical and emotional support of the older person, the shops, restaurants, social activities and so forth that are offered within these new settings also play a significant role in facilitating new social and friendship networks amongst residents (Kontos 1998; Croucher 2006; Evans and Vallelly 2008). Kontos's (1998) work on supported accommodation in the Canadian context, for example, pointed to the way in which residents within such settings can form informal and reciprocal support networks that can be critical to facilitating their independence. In her study, wealthier, but less able, residents were shown to help more able but less affluent residents financially, in return for help with physical tasks they, themselves, were unable to do. Residents not only gained physical and financial support from each other, but the reciprocal relationships that developed also provided social and emotional sustenance. Given that these settings have care 'thresholds' beyond which an older person is 'encouraged' to move on – even if they do not want to (Cutchin 2003), the development of such reciprocal relationships is argued to not only help older people to continue living in the supported care setting, but critically to help them to avoid or delay progression to residential care.

Work in this vein highlights the potential of such sites to become alternative communities of care and cohesion and the basis for social organisation. It also illustrates how the construction of home in these settings is intimately linked to a common struggle to live 'independently'. For the older people in Kontos's (1998) study, 'independence' was linked to the ability to avoid the need for formal care provided by staff or others not part of the supported living environment. Crucially, however, this was gained through a reciprocal *interdependence* between co-residents. The relationship between home and independence within these settings is also linked to the older person having their own 'front door' with all the rights inferred from owning or renting his or her own property (Percival 2001). So despite the transition from the familial home to a supported care setting, by and large, the older person retains the ability to decide who to exclude and who to allow entry and the freedom to make decisions about how he or she organises and lives their life within that home. Indeed, these examples highlight some of the complex ways in which older people accept their increasing frailty whilst actively resisting any attempt to convert the home environment into an institutionalised site of care (Kontos 1998).

Whilst segregated housing options for older people can provide important care alternatives, there is conflicting evidence over the socially beneficial nature of these environments. Indeed, critics have argued that such developments can act to further marginalise and isolate older people from the community (see for example, Fennell 1982; Golant 1999; Percival 2001). Not only do these critiques illustrate how hierarchies of frailty are formed as some residents become frustrated and intolerant of those who are more frail or dependent than themselves, but as Percival's (2001) work demonstrates, new tenants can feel an acute loss of the community and friendship networks they have left behind. Rather than supportive communities, he suggests that older people are likely to find the transition a challenge, 'given the rules of engagement, social cliques and antipathy to outsiders which characterise all small communities' (836). This may be further exacerbated where an individual does not feel stimulated or engaged by the social activities and interaction that take place in communal areas, resulting in loneliness and a sense of isolation.

This sense of marginalisation can be exacerbated by the enforcement of rules and regulations designed to ensure the smooth running of the supported living environment. The tensions that can arise between residents and paid support staff can be interpreted by residents as features of institutional living (Cutchin 2003). For example, conflict between paid staff and residents over the ownership – and consequent control and usage – of communal spaces can be seen by residents to not only compromise their independence, but to threaten their personal identity. Such conflict can be at odds with residents' construction of these places as home, leading older people to actively attempt to resist staff directives in the attempt to negotiate their own terms of existence (Kontos 1998). Such conflict can be seen as a struggle between 'home' and 'institution'.

So while transition to these supported care settings may, on the one hand, be conducive to the development of friendships, informal support networks and community formation in ways that can maximise an older person's ability to avoid the need for institutional care, it can also prove an alienating and disempowering experience. Age-segregated settings can thus be experienced in both positive *and* negative ways according to the effect of the physical and social environment on the older person's sense of self and identity and consequent impacts on social motivation.

It is also worth noting that current availability of Extra Care in the UK is limited. Where demand exceeds supply providers are able to set eligibility criteria that prospective residents have to meet. Such criteria are liable to exclude those most likely to make high demands on providers' time and those whose behaviour is seen as disruptive (e.g. those with dementia). Those desiring to relocate to retirement villages are not only expected to be in relatively good health on entry, but as comparatively expensive places to live, they also need the available financial resources to do so. As Croucher's (2006) work on retirement villages in the UK has amply illustrated, few residents within these settings are eligible for means-tested benefits. So whilst the concept of 'home for life' offered by retirement villages may appear an attractive option for older people, in reality it favours relatively healthy, wealthy older people. This leaves Sheltered Housing and Extra Care as the main housing alternatives to domiciliary care for those less affluent older people who are no longer able or willing to reside in the familial home. But this too raises concerns. Many Sheltered Housing options are not only ageing and in need of modernisation (Evans and Vallelly 2008) but as a recent report by Help the Aged (2009) has highlighted, there is significant evidence that on-site wardens are being replaced with floating support. Indeed, the report estimates that 31 percent of Sheltered Housing schemes in England will lose their on-site wardens between 2009 and 2012. This is not only occurring against residents' wishes, but is raising significant concerns that older people will be left at 'possible risk' from slower response times to emergency calls.

Care, Community and Globalisation

Valuable and important though studies of care, community and ageing in place are, work by Rowles (1993), Phillipson (2007) and others has begun to suggest that the fundamental basis of this work this is being transformed by the social and economic processes associated with globalisation. Understanding the influence of globalisation at a local level, may thus hold the potential to offer important new insights into the relationship between older people and the nature of care, community and place attachment. So, for example, while past generations may have been less mobile during their working lives and hence more attached to particular places than upcoming cohorts, this is likely to change. Evidence is emerging of shifting migration patterns as 'younger-old' people move to warmer or more scenic locations after retirement (the so-called 'snowbird migration') followed by relocation to places closer to their families as they begin to recognise they may require informal care and support as they age (Milligan and Wiles forthcoming). Many individuals currently in their 50s and early 60s also have a work history that is marked by far greater mobility than earlier generations. For these individuals, attachment to a particular place – particularly home, is less likely to be marked by temporal depth and more likely to involve a more diverse array of environments than earlier cohorts. Hence, globalisation and these new types of movement both prior to and within old age, are constructing and expanding the mix of spaces, communities and lifestyle settings inhabited by older people. This poses new questions about older people's sense of belonging to, and integration with, particular communities and environments.

In relation care policies designed around ageing in place, the impacts of the global on the local may be especially important given the length of time individuals are likely to have resided in the same community and the extent to which their mobility may at some point be restricted to defined territorial boundaries (Krause 2004). For some, such changes may be manifest in an expansion of opportunities in respect of residential choice and location – and ultimately where care and support may take place. Others, however, will find themselves relatively disempowered from these options. Phillipson (2007) for example, maintains that these global processes are likely to manifest in the emergence of *new* social divisions, between those able to choose residential locations consistent with their biographies and life histories, and those who experience rejection or marginalisation from their locality. Hence, one view of the impact of globalisation for older people is that it is fragmenting and distorting the experience of community and place for older people. Rather than the spirit of co-operation and mutual solidarity often presented as typical of local communities in the mid-twentieth century, community life is now viewed as unstable, disengaged and friable (Beck 2000). The links between globalisation and issues of rootlessness, mobility and impermanence are thus argued to limit the relevance of community and place for older people.

Such critiques infer that unless policy recognises the impact of globalisation, any attempt to foster ageing in place through efforts to reinvigorate a sense of community that will be supportive and caring of older people are likely to be doomed to failure. Instead, it has been suggested, that globalisation provides an opportunity to re-conceptualise issues relating to community and place in later life. Such an approach is seen to offer the potential to gain new perspectives on care and older people (Phillipson 2007).

Globalisation also introduces complexity at a local level through the emergence of transnational ties and relationships, which themselves may evolve into communities in their own right. As Levitt (2001, 4) suggests, the rise of the transnational community reflects 'how ordinary people are incorporated into the countries that receive them while remaining active in the places they come from'. 'Snowbird migration' for example, brings to the fore issues around dual residence, migrant care workers highlight the importance of transnational ties and so forth.

These developments suggest that different forms of community may be emerging – ones that are more mobile and transient and which give and receive care and support in different ways. Technology is also changing the way we communicate, shop for aids and services and gather information. We no longer need to be in the same town – or even the same continent – to communicate regularly and gain support from friends and family, drawing into focus debates around the 'blurring of the boundaries' between proximity and distance in relation to caring *for*. In addition to the shift toward more mobile populations of care workers and older people, commentators also point to the way in which local environments are, themselves, being transformed by the diverse social, cultural and economic changes associated with globalisation. This is manifest in the emergence of a global culture that is homogenising places (Massey 2002). Many high streets and shopping malls contain the same stores and have a similar feel, hotels and shops have a corporate 'look' etc. – all of which contributes to a desensitisation to place. Critics thus maintain that we are moving inexorably toward a position in which 'every place can be anyplace in an essentially placeless world' (Relph 1976).

Chapter 8
Care and Transition –
From Community to Residential Care

Transition involves critical turning points or events in the lives of individuals (Meleis et al. 2000). Whilst everyone will experience some critical junctures during the lifecourse, older people will experience multiple transitions starting with the transition from work to retirement. With increasing age and frailty these transitions can also include a shift from active to passive, from good to poor health, from care-giver to cared recipient, and ultimately from life to death. Each of these transitions is bound up in a particular spatial expression of ageing and care. This chapter is concerned with transitions in care from home and community to residential care settings. Following a brief discussion of the context of residential care in the UK, it examines how these transitions are negotiated and experienced by older people requiring care and support and their informal carers. It also considers how informal carers construct new caring identities within residential care settings before addressing issues of home and Home, place and non-place and sequestration within residential care settings.

Care and Transition: Residential Care Settings in the UK

Whilst ageing in place and supported care options provide alternatives to care home settings, with increasing frailty comes an increased likelihood of transition to residential care. This can arise either because the need for care and support intensifies beyond the ability of the informal care to cope, or where no informal care is available; the extent of formal care that would be required to support the older person to remain in the community exceeds that which is available. Indeed, despite the shift towards ageing in place Andrews and Phillips (2002) point out that demographic change in the UK, coupled with complex family changes in recent decades, are likely to cause an overall increase in the demand for residential care.

In the UK, all care homes for older people – whether providing residential only or residential and nursing care – are termed 'care homes' and as such, they must be registered with and regulated by the central government Commission for Social Care Inspection (Ford and Smith 2008). Care homes may be owned and operated by private individuals, companies, not-for-profit organisations or local authorities. Whilst they may vary in size and facilities, all provide living accommodation, meals, personal care such as dressing, and supervision of medication, companionship and on-call night cover. Care homes with nursing care provide more intensive medical support for those who may be very frail, bedridden, or who have a condition requiring regular medical attention. Such homes are required to always have a qualified nurse on duty. Individuals with a diagnosis of dementia may need to be [re]located within a care home that has specific provision for what has been termed 'elderly mentally ill' (or EMI).

The post-war history of residential care in the UK is one that has been marked by a shift from largely state to largely private sector provision, and a period of rapid growth from the 1970s to the early 1990s followed by a gradual decline. The state has long had a role in the funding and provision of residential care for older people and indeed was the main provider of such care in the post-war period. This was supplemented by much smaller voluntary and private sector provision. Successive governments, however, have sought to reduce the state's role as a residential care provider through the encouragement of private sector provision. This encouragement combined with guaranteed state support for residents in private care homes and the closure of long-stay hospital wards resulted in massive boom in private sector residential care, making it the largest provider of care homes in the UK (Andrews and Phillips 2002). A 43 percent growth in all care home places available for older people between 1981 and 1991 for example, was due almost entirely to growth in the independent care sector (Laing and Buisson 1992, 156). Subsequently, however, there has been a significant shift in both the level of provision and the structural organisation of long-term institutional care. Since the end of the twentieth century bed numbers have been declining. In the calendar year of 2001, for example, no fewer than 827 care homes (16,600 places) closed in the independent care home sector (Ford and Smith 2008). In large part these closures are attributable to: the withdrawal of guaranteed state support for residential care in the early 1990s; the impact of community care; and the refocusing of policy toward ageing in place. Care homes have since had to compete with each other for a much smaller number of means-tested clients funded by limited local authority budgets. Some research findings also suggest that the tightening of state regulation on residential care standards (including higher environmental standards and space requirements) have been a contributory factor to the closure of at least some homes over this period – particularly where owners felt they would be unable to comply with the new standards (Netten et al. 2005).

State intervention has thus contributed to both the rise and relative decline of private sector residential care in the last few decades. Yet as Andrews and Phillips (2002) point out, substantial state intervention is critical to the development and provision of residential care in the UK. Left to an unregulated market, they argue, the relatively high costs of residential care would prove prohibitive to less affluent older people and their families. Private providers would also be likely to 'cherry-pick' low-dependency, low-cost residents in order to maximise profits. Uncertainty about demand and potential clients' imperfect knowledge about the sector would also leave it open to poor planning and potential exploitation. Hence, without state intervention, the pattern of private provision would be based less on need and more on ability to pay. As a consequence, providers would be likely to seek to locate close to the most affluent areas, contributing to both a social and spatial inequity of provision.

Informal Carers and the Transition to Residential Care

Entry to residential care is usually the option of last resort taken when all other alternatives have been exhausted. Whilst this clearly marks a critical turning point in the life of the older person, it is also important to recognise that these transitions also have a significant impact on the lives of informal carers. Though most informal carers want to care, without adequate support, the pressures of increasing frailty combined with mounting levels of personal and medical care needs can lead to a breakdown in their ability to cope, increasing the likelihood that the care-recipient will enter residential care. This can be a period of extreme stress and confusion for both informal carers and care-recipients.

How the transition occurs can vary. For some, it may arise as a consequence of carer ill-health, or where a period of residential respite is extended to permanent care as the informal care-giver realises he or she is no longer able to continue caring. For those older people who have already made the transition to supported accommodation, it may arise where care needs come to exceed the levels of support the care setting is able or willing to offer and the informal carer is unable to compensate. In many instances the transition is preceded by an acute incident resulting in the hospitalisation of the care-recipient. Common reasons for hospitalisation include falls within the home resulting in broken bones (commonly a hip), infections and strokes (Milligan 2006). Alternatively (and particularly in cases of dementia) a care-recipient may be hospitalised for a period of assessment (usually over a period of several weeks). For informal carers this can be an extremely difficult time as they are not only worried and anxious about the decline in health of the care-recipient, but also exhausted from their care-giving role. How health professionals in these settings respond to, and communicate with, informal carers at this stage is of critical importance to their care-giving experience. For some, this can be a very positive event, particularly where hospital and allied health staff work effectively as a team and clearly engage with family and informal care-givers in positive and productive ways. Ensuring that medical staff provide clear and understandable explanations of the future care needs of the care-recipient and what this is likely to entail is seen as critical to improving informal carer understanding. Hospital social workers also have an important role to play in clearly explaining the options available to meet the care needs of the patient, the processes that need to be put in place to achieve this and what this will, in reality, mean for both carer and care-recipient.

Not all informal carers, however, have positive experiences at this stage of the care transition. Some note that in general hospital settings doctors and consultants can fail to fully explain the outcome of an assessment and what this means in terms of the options for care. Nor do they always fully appreciate the impact this will have on both the older person and their informal carer. As articulated by one carer in a New Zealand-based study, there is a need for health professionals in such settings to not only ensure that the language and terminology used is easily understandable to lay people, but to recognise the extent of distress that a recommendation to transfer care to a residential setting can have on both the older person and their informal carer. As one informal carer commented, 'the second needs assessment was carried out by a bullying doctor who I found to be most unhelpful. I could not understand what he was saying and was in a state of shock about the whole episode.' Carers have also noted how staff in general medical settings can have limited understanding of how best to work with and relate to older patients. This can lead to a tendency to infantilise or objectify the care-recipient. Indeed, one spouse carer was lead to comment, 'its just the *strange* way they [medical staff] treated him, like he wasn't a human being or somebody with a bit of intelligence, he was just "that thing in the bed."'

Given the ageing of the population and that the incidence rate of frailty and dementia amongst the oldest old is likely to increase (Wanless 2006), it is clearly important that medical staff located in general hospital settings have a clear understanding of how to communicate effectively with frail older patients, especially those with cognitive impairments as well as an increased knowledge of how to manage their specific care needs.

Choice and access

Identifying who makes the decision about entry to residential care is not straightforward. The literature on care home transitions, however, indicates that informal carers play a key role in both the initial decision to seek residential care and selecting a home, as well as completing the necessary paperwork and arranging finances to cover the costs of care (Arksey and Glendinning 2005).

Whatever the reason for entry to residential care, issues of choice and access to the new care setting are critical. Even the briefest of searches reveals that there are currently as many as 21,500 residential care homes in the UK (www.carehome. co.uk). Yet while carers and care-recipients may be presented with what at first glance, appears to be a bewildering assortment of choices, the reality is that once issues of access, available bed-space, ability and willingness of the care home to meet the particular needs of the care-recipient, as well as personal preferences and prejudices have been taken into account, choice is often severely limited. This can be even more problematic where the older person is no longer mobile or has cognitive or behavioural difficulties (for example, the wandering and aggressiveness that can be symptomatic of some stages of dementia). Hence as Arksey and Glendinning (2005) point out, in reality, choice over care home entry does not have the positive, empowering implications or connotations suggested by current policies.

Ease of access in order to maintain regular contact with the care-recipient is a critical factor in the choice of care home. Most informal carers are concerned that the care home should be either within walking distance, close to a regular bus route or require only a short car journey and have ample parking. Ensuring that the care-recipient is located in surroundings that are familiar, feel safe and hold the potential to maintain existing social networks where possible is also important. But limited choice, combined with the informal carer's inability to continue caring at home means that major decisions affecting both themselves and the care-recipient are often made at a time of crisis, when the informal carer may feel pressured, stressed and unsure of the options available. It is interesting, however, that despite these constraints, there is little evidence of 'shopping around' once an initial placement has been made. Indeed, even where disillusionment arises as a result of the care home not providing anticipated services or the level or quality of care expected, few informal carers seek alternative placements. Relocation is deemed difficult if not impossible, in part due to the distress this is likely to cause the care-recipient; in part due to perceived difficulties in dealing with the search for alternative accommodation; and in part because most residential places in the UK are part-funded by the state and hence would involve engaging with another layer of bureaucracy (Davis and Nolan 2004).

Entry to care homes: Knowledge and information

For most informal carers, the entry of their spouse or close family member to residential care will be the first time that they will have had close contact with a care home. Understanding their role in this process, what they can expect from the care home, what a care home might expect from them, as well as their rights and responsibilities with regard to care can be very confusing.

Informal carers gain knowledge and information about care home processes from a variety of sources. Some may have a good understanding of the process and options due to their previous life experiences – either from having worked in the 'industry' or having been involved in arranging residential care for another relative or close family member at some point in their life. Others actively seek information from published and publicly accessible sources such as books and the internet, whilst others draw on the experiences of friends and relatives. Many, however, have little idea about where they can gain knowledge and information, and may receive little more than a list of care home that will be reduced to those with available beds. This basic lack of information about care homes and the processes of transition can add to the worries and concerns that informal carers hold about the future for both themselves and the care-recipient.

Most care homes offer opportunities for informal carers to visit and view the home prior to making any decision about entry. In some cases health professionals or social care staff encourage informal carers and care-recipients to undertake these visits some time prior to the point of transition in order to inform themselves about future options. The benefit of taking such a proactive approach is reflected in the following carer's comment:

I didn't know *anything* about care homes or anything like that, so I think its better if you *do* know, and you go round them if you can, and sort something out *early* and kind of get to know that Home before they go in. You know, sort of visit and get to know the people who are working there, what its like and what's expected of you and what you expect of them.

Yet as Davis and Nolan (2004) also point out the ability to plan proactively is limited as many entries to residential care are the outcome of care crises. As a consequence, even where a positive choice is made, there can be no guarantee that a place will be available at the point of transition.

For others, visiting a care home prior to the stage of entry can prove a difficult or depressing experience. In part, this may be linked to the care home/s visited, for example where décor may be old and depressing, where facilities are limited, or the physical layout overly institutional. In addition, the frailty of existing residents can act to remind the informal carer and care-recipient of the extent of decline that can be anticipated. Transition to residential care is marked by the fact that neither care-recipient nor informal carer can look forward to recovery; rather they have to face decline as an ongoing process. Care homes also face a dichotomy between presenting neatness, and thus meeting the expectations of informal carers, and revealing the untidiness and disorder that can arise where the ethos of the care home focuses on maximising the independence of the care-recipient (e.g. through encouraging them to undertake tasks such as dressing, taking care of their own personal spaces, basic housekeeping tasks etc.). Care home managers also note that whilst informal carers view the care home in the light of what they believe the care-recipient would most like, the tendency is to view this through the mindset of what the care-recipient was like prior to the onset of frailty rather than through the parameters of their current care needs. Of critical importance here, is the extent to which care staff introducing the informal carer to the care home recognise this lack of understanding and encourage the informal carer to view the setting through the eyes of the care-recipient. At the same time, care home managers need to have a clear understanding of what aspects of residential care informal carers and care-recipients perceive to be important in providing a home-like setting.

Care homes can appear a strange and alien environment for both the care-recipient and the informal carer, hence it is important to make these settings as familiar as possible for both whilst establishing good relationships with formal care staff (Davis and Nolan 2004). Where a place is available in a care home that is already known to both carer and care-recipient this can substantially ease the process. As the following carer noted:

When we arrived at [care home] the staff came out to greet us and help us. They were very welcoming and went out of their way to make us both feel at ease. I was invited to stay for lunch, which I did, and that gave us both plenty of time together before I left. As [spouse] had stayed there before and actually was given the same room he was in previously, he already knew many of the staff and residents. That made it much easier for him.

Care transitions over time

As noted at the beginning of this chapter, older people can experience multiple transitions during their old age; few studies however, have addressed the experience of multiple transitions on informal carers. Those that do exist (e.g. Bowers 1988; Ryan and Scullion 2000a 2000b; Hertzberg et al. 2001; Milligan 2005) indicate that while both carer and care-recipient can experience considerable stress and anxiety during the initial transition to a residential care setting, contrary to expectations, this anxiety does not always lessen over time.

Most care homes insist on an initial trial period in order to assess the suitability of the potential resident. Where a care-recipient is judged as too disruptive or requiring greater support than that offered by the care home (as with supported accommodation) this can lead to a request that an alternative placement for the care-recipient be sought leading to a further move within the first few weeks of transition. Yet considerable distress can also be experienced where subsequent behavioural patterns or changes in the health and care needs of the care-recipient following initial admittance mean that the care home is no longer willing or able to provide support for the care-recipient. In such situations, informal care-givers can find themselves in the position of having to find alternative care home accommodation as well as having to explain the need for the shift to a care-recipient who may just have begun to come to terms with the first transition.

The requirement to find an alternative care setting, however, can also arise as a consequence of: the informal carer's failure to recognise the extent of decline in the care-recipient's abilities, and hence his or her unrealistic expectations about what constitutes a 'suitable placement'; and the care home's ability to provide for the specific care-recipient's needs. As one carer put it, 'with the progression, well you never keep up with it – you never want to be where it's at – it [declining health] always progresses faster than you actually can accept it.' Inappropriate placement can also arise as a lack of clear communication between those making the care assessment, the prospective care setting and the informal carer about what actual level of support a care home will need to provide for the care-recipient as well as any known behavioural problems.

Not only do these transitions act to reinforce the increasing frailty of the care-recipient, but where that person has been resident for some while, the build-up of trust and communication that may have developed within the previous site of care is lost, leaving both carer and care-recipient to develop new relationships with paid care staff within the new site of care. For the older person, this can also result in the loss of any new friendship networks they may have begun to develop within the new care settings. Secondary transitions of this kind can be particularly prevalent in smaller care homes that have limited capacity to deal with more intensive levels of care. Many larger care homes, however, now have separate wings designed to cope with differing levels of frailty or behavioural issues. As a consequence, as care and support needs change, the shift may be to a different wing or floor of the care home rather than a different home altogether. Nevertheless, many of the issues referred to above apply equally to moves even within a care home.

Informal Carers: Constructing New Identities in Residential Care Settings

Evidence concerning informal carers' continued contact with the care-recipient following transition to the care home is mixed. Whilst some researchers maintain that the extent of informal carer contact will decline over time, others reveal that in fact many spouse carers continue to visit on a regular (often daily) basis and many adult child carers visit several times a week. Furthermore, many of these informal carers continue to actively participate in the care and support of the care-recipient despite entry to a residential setting (Belgrave and Brown 1997; Milligan 2005). Many informal carers will have been heavily involved in the care of the older person prior to transition and as a consequence their identities become intimately bound up with 'being a carer'. Their changing role in the new care setting means that the informal carer may have to construct a new caring identity for him or her-self within that setting. This reinforces the fact that informal carers do not necessarily revert to the role of 'visitor' following the care transition, but in fact can continue to play an active role in care-giving to their spouse or close family member. Indeed, the level of informal contribution to the care of older people in care homes can be surprisingly high.

In seeking to retain identities as individuals who both care *for* and *about* the care-recipient, informal carers can undertake a range of caring tasks within care homes. These can be seen to fall into four main groups. Firstly, they can engage in a range of physical care tasks such as feeding, limited personal care (cutting nails, occasional washing, changing of soiled and dirty clothing etc.), laundering clothing, purchasing personal equipment, providing toiletries, clothing and 'treats' such as flowers or sweets and personalising rooms. As one informal carer put it:

> Well, I feed him and given him his drink, and you know, is this the right food or would you rather have something else? And little personal things, and I would have his radio in there. *Oh, yes, he was still very much my 'Tim'* ... little everyday things like feeding and taking him treats and things like that (Milligan 2006, 325).

As the above interview excerpt also reveals, here the spousal carer is not only constructing a new caring identity for herself, but is reaffirming the strength of the spousal link and hence her 'right to care' despite the transition from the domestic home to residential care.

Secondly, informal carers can undertake a range of social tasks, such as visiting and entertaining within the care home, accompanying the care-recipient on walks, taking him or her on outings, drives in the car, or to family events. They also act as a conduit to the outside world by keeping the care-recipient informed of local and family events. Thirdly, they can undertake a crucial monitoring role, checking on the quality of care the care-recipient receives as well as monitoring treatment, medication and personal appearance. Where problems arise they can ensure that these are raised with the appropriate care staff with aim of ensuring that issues are satisfactorily resolved. This monitoring role is, of course, primarily directed at their own spouse or close family member, but research has shown that some informal care-givers also take on a secondary role of monitoring and supporting other residents. This is particularly directed at those who have few visitors – either by regularly spending time visiting and befriending them, or by also monitoring the quality of their care. As one informal carer put it:

> ... not everybody has got family, so what I do is have one or two other people that I keep contact with when I go in. You know, pick up on somebody that doesn't seem to get visitors and that sort of thing and just make a point of saying hello, and taking them for the odd walk and whatnot (Milligan 2005, 2116).

Fourthly, they engage in the emotional work of caring *about* by demonstrating affection and love to the care-recipient, and by providing companionship and emotional support. As one spousal carer put it: 'I make a point, with my wife, when I do see her, I always give her a cuddle and I usually smooth her hair or whatever.' While the emotional work of caring may be less frequently referred to by informal carers it underpins many of the other tasks undertaken. For example, informal carers would be unlikely to engage in the wide range of monitoring, physical and social care tasks they undertake, unless a deep bond of love and affection existed between them and the care-recipient.

Of course not all informal carers will want to continue to remain actively involved in the day-to-day care of the care-recipient, but where they do, they can often find themselves thwarted by rigid and insensitive practices within the care home. The well-meant, but somewhat tactless advice not to visit for a few days in order to let the care-recipient 'settle in' that is often given to informal carers on first entry to the care home, for example, sends an immediate message that the informal carer is no longer expected to play a significant role in the care of the care-recipient (Davis and Nolan 2004). Here, informal carers are often typecast as 'visitors' by care staff who fail to recognise – and thus benefit from – their expertise and unique care-giving histories. Yet informal carers carry a wealth of knowledge and expertise about the health and well-being of the care-recipient, their likes and dislikes, interests, patterns of behaviour and so forth that can be invaluable in improving the quality of care (Milligan 2006). Indeed, Wiles (2003, 10) goes as far as to suggest that, 'Informal caregivers are often so attuned to the needs of the person for whom they care that they can predict crises before they occur and mitigate them, their contextually specific skills thus balancing the professional expertise of nurses and formal workers.' Where care staff fail to acknowledge or engage with this expertise, informal carers are often left to find their own ways of ensuring their skills are employed. Researchers, for example, have pointed to the adoption of covert tactics to monitor the care, health and well-being of the older person and ensure they are settling in (Davis and Nolan 2004; Milligan 2004). Covert tactics can also be adopted where informal carers are made to feel that they are intruding or 'stepping on the toes' of the 'professional carer', even when undertaking some of the most basic of aspects of care such a assisting with meals, small personal hygiene tasks and so forth. As one informal carer explained, 'We weren't encouraged to do anything in the personal hygiene area – although I always continued to cut her toenails and fingernails – although I always waited until there was nobody around before I did it' (Milligan 2004, 58).

The interconnectedness between formal and informal care-giving thus adds a further dimension to the caring relationship – one that is concerned with conflict. As Tronto (1993, 109) notes, while ideally care involves a smooth interconnectedness between the two, in reality there is likely to be conflict between them. The views of formal and informal carers may vary rendering notions of partnership working between formal and informal care problematic particularly where care 'professionals' assume an 'expert' knowledge that is at odds with the deeper personal knowledge of how best to deliver care for the individual held by the informal carer. Secondly, conflict can arise between different levels of the formal care-giving institution. Often in bureaucracies, those who determine how needs will be met are far removed from the actual care-giving and care-receiving, resulting in conflict about how that care can best be delivered and resulting in a poorer quality of care. This can be particularly so where managerialist values clash with the ethos of care held by nursing and care staff. Care-recipients may also have different ideas about their needs than do care-givers and may want to direct rather than be passive recipients of care. All provide a potential recipe for conflict that needs to be resolved if effective partnership working is to be constructed.

So while care staff are often highly caring and concerned individuals, they can nevertheless be constrained both by the bureaucracy of the institution and the need to balance work-based rules and regulations against the needs of older people residing within the care home – each of whom will have varying needs and wants. Inevitably this will limit the time and resources they have to spend on any one individual, leading to sometimes unrealistic expectations and the potential for conflict between informal carers and care staff. As one care home manager put it when asked whether she felt people's expectations may sometimes be too high:

> I think so. I mean oh, it's the same in any hospital or any home like this, you can't have it *as* you have it at home you've got to give and take a little bit. Not to the detriment of the patient or that, but just to have a little bit of understanding sometime. The staff are human too and they can't do the impossible! ... And I think *some* of them [informal carers] are really quite unreasonable (Milligan 2004, 55).

Indubitably care homes should ensure the delivery of high quality personal and (where relevant) medical care in a safe and warm environment. Their inability to provide one-to-one care, however, means that inevitably they are unable to deliver the levels of personal care an older person is likely to have received from the informal carer prior to entry into residential care. Further, while formal carers have the technical skills to care, they lack the biographical expertise that would most effectively help them to carry out those technical procedures of care (Ryan and Scullion 2000a). Hence, as suggested above, good-quality care requires informal and formal care workers to collaborate in the delivery of care and in mutual education between informal carer and care worker.

home or Home?

Crossing the threshold from community to institutional care equates not just to a physical but also a symbolic change in the place of care, one that has been characterised as a move from being [at] 'home' to [in a] 'Home'. Utilising the adjunct 'Home' in relation to residential care has raised significant debate about the extent to which care Homes can truly achieve a 'sense of home' (see for example, Groger 1996; Nolan and Dellasega 1999; Peace and Holland 2001; Milligan 2003). As discussed in Chapter 5, 'home' is wrapped up a sense of safety, identity and self-expression that reflects an individual's history and preferences. Such attributes facilitate both a preconscious sense of self and the ability to maintain control of ones life, promulgating independence. Care in the home, for example, means that the care-recipient still retains some control over what things are actually done, how and when. In residential care Homes, however, the customary routines and practices of the individual do not always suit those of the institution – for example, routines around when to get up and go to bed, when and what to eat, when to bathe and when not. As a consequence the resident's routines may become subsumed by those of the institution and its workforce. Further, as semi-public places, care homes are open to official scrutiny. Whilst such scrutiny is designed to ensure the protection of vulnerable people, it is also important that those in residential care are able to take reasonable risks and have a degree of autonomy (for example around decisions to undertake certain activities within the Home – or indeed to undertake activities beyond the scrutiny of the Home). As Peace and Holland (2001) point out, there is a tendency for social policies directed at those in residential settings to be paternalistic and overly protective in a way that far outstrips that likely to be experienced in the domestic setting. The constraints of communal living, combined with the divergent roles of residential care as both sites of 'home' and work, means that care Homes expose the tensions between public and private, empowerment and disempowerment and between domestic and institutional living.

In seeking to make care Homes more 'home-like', most encourage care-recipients, informal carers and family to bring in small items such as pictures, photographs, ornaments and occasionally a favourite chair and so forth, in an attempt to imbue the private space of the bedroom with a feeling of home and identity. But it would of course be impossible for this to be carried out through all communal rooms. The care Home may thus look 'homely', but the key question is, whose home? Indeed, most communal areas within care Homes reflect the proprietor's tastes and design requirements and are 'governed' by their [or the care manager's] view of what is appropriate activity and behaviour. The desire to provide a home-like environment is thus in tension with the need for practical furnishings (e.g. waterproof and wipe-clean soft furnishings) and the demands of regulatory health and safety requirements (Peace and Holland 2001). Additionally there are still a few care Homes where residents must share bedrooms and thus have no private space to call their own. Even where individuals do have their own rooms, where residents have some form of cognitive impairment the lack of locks on bedroom doors means that they can regularly wander in and out of each others rooms, appropriating items of clothing, ornaments or other personal items. As one carer commented:

> we were encouraged to bring in pictures, knick-knacks and so on for their rooms, but other residents would wander into his room and take them, and they would open up his drawers and take his clothing ... there didn't seem to be any restraint on patients wandering in and out of each other's rooms. There weren't any locks on the doors, in fact one time one of the other patients locked a members of staff in the toilets so they took the locks of the toilet doors after that (Milligan 2004, 70).

Removing locks from internal spaces occupied by elderly residents is viewed as an act of safety and protection, allowing care staff to enter a room unimpeded, in order to undertake necessary care work. Yet privacy may be a key factor in helping an older person to retain their sense of self identity and control over his or her own body within the residential care setting (Peace and Holland 2001). The tension here is that without constant surveillance, it becomes difficult to discourage residents, particularly those with cognitive impairments, from also entering the only private space another resident has within the Home. Further, the requirements of work (through easy access to residents who need care and support) are brought into tension with the resident's need for privacy and the ability to exclude. Current technologies, in the from of key-coded entry panels, can to some extent resolve this dilemma – allowing staff entry whilst excluding other residents, but it still relies heavily on formal carers respecting the resident's desire not to allow entry – a wish that is often overturned as the need to perform care work and respect privacy are brought into tension.

This need for private space, however, depends to some extent on the level of frailty and cognitive ability the care-recipient experiences. For those who have a good level of awareness and are able to converse, the protection of private space can be critically important to the experience of 'good care' in the Home environment. But where the care-recipient is no longer able to respond to conversation or has very limited mobility, it can be more important for him or her to spend time in communal spaces where they can see and hear the everyday activities of the care Home.

Inevitably an individual's social life alters in the move to residential care. For some, this can mark a shift from isolation at home, and as a consequence they may enjoy the ebb and flow of the daily life and routines of the care Home. Many, however, find themselves cut off from their previous social networks. For both, life becomes centred around the Home. Interestingly, however, Peace and Holland's (2001) work on care Homes in the UK found that whilst residents occupied communal spaces, not only did they not 'visit' each other, but they rarely engaged in activities together unless it was an activity organised by care staff. Residents appeared to protect their privacy and deal with circumstances they found annoying within the Home by avoiding intimacy with others. Indeed, they note that 'on the whole the atmosphere between residents might best be described as a polite interest or tolerance, rather than close friendship' (404). This perhaps reflects the homogeneity of older residents who live together not through choice but necessity.

Whose Home?

Care Home size may be an important factor in the extent to which they can generate a 'sense of home'. That is, that smaller, more intimate, care settings may enhance the ability to develop closer caring relationships between formal and informal carers, facilitating the development of more integrative care. Here, is possible to distinguish between 'extended family homes', that is, small Homes that are also the proprietor's own home and 'mini-residential Homes' where the proprietor lives elsewhere (Peace and Holland 2001). This raises the crucial issue of whose home? In the former setting, residents and proprietors share the same home. Such an intimate sharing of living space offers limited options for older residents to create either private moments or private space. Indeed, Peace and Holland (2001) noted that whilst the proprietors of such small Homes are keen to ensure that regulations and good care practices for their residents are upheld, this inevitably creates tensions between place as work and place as home – but in this case tension refers to the desire to retain a family-like home for their *own* families. Where tension arises, the proprietor's needs take priority over those of the resident. Drawing on Nippert-Eng (1996), they point out that, 'work', as a public activity, requires the public presentation of a carefully constructed self. Appearance, speech, emotions, and the portrayal of intellect 'at work' is attended to in specific, situationally defined ways. As a private realm, however 'home' is the place where we can 'be ourselves', 'put up our feet', 'let down our hair', relax among those who see us 'warts and all' but aren't supposed to hold it against us (Nippert-Eng 1996, 20). In the care setting, where the proprietor or care home manager lives out, the distinction between residents and carers is clearer, and the nature of the place as a residential care Home rather than a family home more obvious.

Addressing the issue of 'whose home' also highlights the tension between those facets of a care Home that the informal carer may feel are important in the construction of 'home' and those important to the care-recipient (who may deny or resist the need for care Home entry). For many informal carers, the physical entity of home can be seen as secondary to the affective aspects of home. That is, the extent to which the atmosphere within the care Home engenders a sense of safety, love and caring – those affective feelings of warmth and inclusion that stem in large part from the interrelationships that develop between informal carers, care-recipients and the care Home staff themselves. As one carer explained:

> I think it's about making it into a home, making you welcome and part of the place – your not just a visitor ... feeling comfortable and at home in a place is better than what colour of wallpaper you've got. I mean it doesn't really matter too much about the building (Milligan 2005, 2118).

Conversely, exclusionary and institutionalising facets of the care Homes can be reinforced by being made to feel unwelcome by formal care workers and excluded from participating in, or decision-making around the resident's care.

The ideal put forward in relation to residential care settings is still one in which the domestic or familial model of 'home' is still central. There are perhaps three key dimensions inherent within this concept of 'home':

- the physical, which incorporates objects and defines boundaries and spaces (and further endows the individual with the power to exclude);
- the social – involving relationships between people and their interactions; and
- the emotional, where the attachment to home is associated with feelings of safety, identity and meaning (Milligan 2003).

Care Homes are thus faced with the dichotomy of attempting to create a homely environment in places that must also be designed to accommodate a diverse range of people with differing work, care and support needs. But it is clear that the tensions between work and home and between public and private make it difficult to conceive of how even the 'private space' of the bedroom can achieve a sense of 'home'. This is inevitably exacerbated by conflicting requirements and notions of what makes a home and whose home is being constructed. Given every older person's own home will reflect their own individuality and sense of identity, and that this may well be in tension with the informal carer's and Home owner's perceptions of what is required to make the Home 'home-like', it seems impossible to resolve these issues. Indeed, some informal carers have pointed out that it may be more appropriate for care Homes to focus less on attempting to achieve a sense of home, and instead recognise that such places have more in common with hotel or motel settings. Indeed, some large 'chain' care Homes are owned by companies also owning hotels. The reception areas and interior décor of these care homes often bear a startling resemblance to that of a hotel setting. Implicit in any equation of care Homes with hotels is the notion of transience – that is, that residents comprise non-stable communities made up of temporary individuals who are merely passing through (in this case in the transition from life to death). As such, while these settings are designed to make the resident as comfortable as possible, and the workforce has a key role in ensuring this occurs, the resident can only ever be seen in terms of a visitor – albeit one for whom care staff may develop an affectionate – but working – relationship.

From Place to Non-place?

The care transition can thus be seen as a shift from a *sense* of home to a sense of Home with all that entails in terms of potential loss of identity, sense of self and preconscious sense of setting. This move from home to Home can also be viewed through the lens of displacement. Here displacement refers to the disruption or disturbance of the normal connection between the informal carer and care-recipient and between those caring relationships that existed within the domestic home and those that emerge within care Home settings. Ryan and Scullion's (2000a 2000b) work on care transitions, for example, illustrated how, as the place of care shifts from the private space of the home to the semi-public space of the care Home, both informal carers and care-recipients can experience a sense of dislocation and a loss of control over the giving and receiving of care. Firstly this occurs through the loss of power to exclude; and secondly, through a loss of power to make decisions about the division of labour between the informal carer and the care professional. As noted above, in general informal carers are expected to relinquish the act of caring *for* to the bureaucracy of the institution. In adjusting to this new role, informal carers and care-recipients point to a range of disempowering experiences, particularly in relation to decision-making around issues such as medication, the daily rhythms of the care-recipient's life within the care Home, personal care plans and informal carers rights to care within the new care setting. Disempowerment can also manifest through the attitude of some staff toward the care-recipient (e.g. infantilisation and loss of 'adult status').

The essentially 'alien' environment of the care Home can make the giving of care a stressful experience for the informal carer. As Allen and Crow (1989) argued, such places cannot in reality be viewed as homes at all, the continued presence of formal carers in these institutional settings, as opposed to the limited time they spend within the domestic home, makes it difficult for both informal carer and care-recipient to establish either private time or private space. As with supported accommodation, communal spaces are not viewed as belonging to or controlled by residents but to staff. Long corridors and bathing rooms with regularly placed antiseptic hand-wash dispensers, the presence of the medical paraphernalia of care such as locked medicine trolleys, oxygen tanks, bath hoists, wheelchairs, formal and 'informal' nursing stations etc. all reinforce this sense of dislocation.

Significant areas of the Home are also 'off-limits' to residents and their families. In turn, the care-recipient and his or her informal carer and family have limited ability to establish spatial exclusion even within those spaces deemed to be private to them. Whilst these spaces are in effect 'rented', unlike supported accommodation, there is no 'front door' nor does the care-recipient accrue any 'tenant's rights'. The application of the term 'home' is thus something of a misnomer and the power balance that is evident within the private space of the domestic home is reversed.

The transition to long-term care can, thus, bring with it a sense of displacement and a loss of identity, indicating a need for both informal carers and care-recipients to renegotiate their identity within the new (sometimes shifting) boundaries of institutional care. For Augé (1995), such sites are characterised by their relative anonymity and as such, represent non-space, sites that lack anthropological meaning for those located within them. Much like the hotel setting, individuals located in the care Home can be viewed as fellow residents placed together through circumstance rather than volition. Residents have little say in who else will reside with them and as a consequence often have little in common with their fellow residents other than their frailty.

Such places then, are 'passed through', their uniformity of objects and rooms and their lack of distinguishing personal features rendering them 'placeless spaces' (Twigg 2000, 78) that are sharply distinct from the 'anthropological place' of the home. Within larger care Homes, residents often pass through a series of wings (or wards) as their condition deteriorates. Relationships formed with both staff and other residents can thus be relatively short-lived. So whilst much policy emphasis is placed on 'continuity of care', in reality this means that residents are moved into different wings designed to deal with differing levels of frailty. As such, they can be seen as 'non-places', sites that lack the connection, memories and identification with place that is characteristic of home. On the one hand, this re-ordering of care can be viewed as serving to provide the best possible care setting, for the individual as their frailty increases, on the other as been argued to be is indicative of a sequestration between life and death (Lawton 1998).

Given the difficulty of realising home within care Home settings, Groger (1996) suggests that this leaves us with two options: either home becomes conceived of as a metaphor for independence and health – and in doing so become irretrievably lost within residential care settings; or home becomes defined in terms of family and social relationships. In this way, she suggests it may be possible to create a feeling of home as long as these relationships continue and are encouraged to continue within in the care Home.

Care and Sequestration

At a symbolic level the transition from home to Home has been argued by some to represent a form of sequestration between the 'fit' and 'frail' and between the living and dying. This perhaps represents the most extreme end of the spectrum of care and brings to the fore a potential conflict between care Homes as places where life is lived, and places where death is regularly encountered (Froggatt 2001).

Those working around ideas of sequestration and care point not only to sequestration between the fit and frail; but also to social sequestration or 'social death', manifest in both the setting apart of the individual from a social perspective and the cessation of the individual as an active agent in other people's lives (Sweeting and Gilhooly 1997; Lawton 1998; Froggatt 2001). Sweeting and Gilhooly point to three groups of individuals who, they maintain can be seen as socially dead within society: very frail older people; those in the final stage of a long-term chronic illness; and those who are perceived to have lost their personhood. Here then, social sequestration is linked not only to bodily decline but also the physical separation (or institutionalisation) of the frail older person, distancing them from those social networks that had previously formed an important part of their everyday lives and identity. At a societal level, Froggat (2001) maintains that care Homes are imputed with the task of containing the visible manifestations of ageing by society, offering a form of social control over those (older) people whose bodies are decaying and as a consequence are no longer able to live their lives without formal care and support. This 'social death' is also manifest in the anticipatory grief experienced by the informal carer, friends and relatives. As will be discussed in Chapter 9 for example, informal carers – particularly spousal carers – often experience the transition of care from home to Home as a period of grief and loss akin to bereavement.

Within care Homes, commentators have pointed to the spatial separation of the living and the dying (Hockey 1990; Froggatt 2001). As already noted, larger care Homes often segregate residents on the basis of frailty. In this way, dying older people are kept apart from those who are perceived to be living (Hockey 1990). Smaller Homes catering to 'fitter frail older people' often request that the care-recipient be transferred to an institution catering for more intensive care needs as the older person's physical and/or mental abilities decline. Hence, even within and between care Homes, the difference between relatively fit and frail is spatially manifest. In part this is also expressed through the different social roles residents come to play within the Home and their shifting patterns of presence or absence within communal and private spaces. That is, as dependency and the need for nursing care increases, individuals whose condition is deteriorating, or who are perceived by staff to be confused, take on a different social role within the care home (Peace and Holland 2001). This is manifest in a decreasing presence in communal areas, declining participation in organised activities and an increasing sequestration within their own bedroom. The spatial setting apart of those who bear the most visible indicators of the processes of dying are thus indicative of the ways in which society copes with the reality of ageing as an antecedent to death (Froggatt 2001).

Lawton (1998) further suggested that identity and selfhood of the individual in contemporary western society is fundamentally bound up with the possession of a physically bounded body. Sequestration for her is thus linked not just to the loss of personhood and identity, but also to the deteriorating and 'unbounded body' through which fluids and matter normally contained within the individual's body are leaked and emitted to the outside, often in an uncontrolled and *ad hoc* fashion. For example, amongst the most frail older people symptoms of urinary and faecal incontinence, ulcerous or weeping limbs and so forth are not uncommon. Care Homes thus represent spaces where the bodily evidence of sickness and death are set apart. Indeed, Lawton maintains that on one level care Homes can be seen to serve as liminal spaces that buffer very frail older people who are wavering 'between two worlds'. In this way sequestration can be seen to reinforce the conceptualisation of care Homes as a 'non-places'. That is, through the operationalisation of spaces within which the processes of bodily deterioration are sequestered, the care home enables certain ideas about 'living', personhood and the hygienic, sanitised, bounded body to be symbolically enforced and maintained.

Concluding Comments

This chapter has sought to engage with debates around the transition of care from domestic and community settings to residential care. In doing so, it has drawn attention not only to how these transitions are experienced by informal carers, but how they seek to reconstruct new identities for themselves within the new care setting. It has also drawn attention to some of the critiques surrounding residential care and how these might be addressed through more integrated care practices (a point I return to in Chapter 10). Critiques are threefold. The first strand focuses on the inability of communal living environments, such as care Homes to provide independence and choice for their residents and to protect their civil liberties. Care Homes have been criticised for reflecting institutional patterns of provision that are concerned as much with custody and sequestration as care. A second strand of criticism maintains that no matter the extent to which care standards are improved, residential care Homes will always have negative associations that not only make them the 'option of last resort' but which act to further marginalise and exclude older people. Third, critics have pointed to the 'home' versus 'Home' debate – arguing that no matter how much care 'Homes' seek to provide homely settings, care regulation, the home/work dichotomy and the structured dependence inherent within care homes will mitigate against their ability to seen as 'home' by their older residents. Within this chapter these debates have been underpinned by an engagement with the shifting landscape of care – more specifically as it pertains to home and Home, and how this might be understood through the concepts of place and non-place.

Chapter 9

Emotion and the Socio-spatial Mediation of Care

Emotions are an integral part of our daily lives, and as such are also a critical facet of the complex landscapes of care. Emotions affect the way we see, hear, touch and respond to the environment, people and places that make up that landscape. Hence, our sense of who and what we are is continually being shaped and reshaped by how we feel (Davidson and Milligan 2004). The interrelationship between the physical and affective aspects of caring and how this both shapes and is shaped by the social and spatial environment within which care takes place is thus important in helping us to understand how care manifests in different ways in different places. Indeed, it has been claimed that for many informal carers, the physical nature and tasks of care-giving, however difficult, are secondary to the bigger emotional impact (Wiles 2003).

Interestingly, few have explicitly addressed the link between emotion, place and care. Perhaps of most significance is Twigg's (2000) work on: *Bathing – The Body and Community Care* in which she highlights the emotional labour of caring for frail older and disabled people in the home. This work draws attention to some of the ways in which care work becomes wrapped up in multiple meanings of home and identity as it increasingly transgresses the normal boundaries of daily life and the public world of service provision. Brown's (2003, 835) work on home hospice and the place of death also draws attention to the emotio-spatial paradox of home/care and its politicisation. Indeed, he maintains that the emotive and material facets of care mark it out as a form of 'heart politics' that 'fracture justice and care in a binary system of political action.' Within this framework, then, attempts to understand emotion or make sense of space can be seen as inextricably linked to care. So while the physical work of care-giving is highly important, work with informal care-givers reveals not only the importance of understanding the emotional impacts of caring, but how this is both socially and spatially mediated (Milligan 2001 2005; Hirst 2003; Wiles 2003).

The Health and Emotional Well-being Costs of Caring

There is increasing recognition that the actual performance of material and affective care contributes to variations in the health and emotional well-being of those involved in that care. In part, this stems from literature on 'expressed emotion' (or EE as it is referred to) drawn from the field of clinical psychology (see Weardon et al. 2000). Research around EE was first developed in the 1950s and 1960s by clinical psychologists as a means of measuring key aspects of interpersonal relationships. In particular this work focused on the impact of family members' attitudes on the relapse rates of people living with poor mental health in the community (Brown 1985). Work in this vein has since been extended to consider other health conditions including the impact of informal carers' expressed emotion on the care-recipient (e.g. Gilhooly and Whittick 1989; Vitaliano et al. 1993; Weardon et al. 2000). The primary aim of this work has been to examine the extent to which aspects of the familial and social environment affect the course or severity of an illness as well as the extent to which the illness affects informal carers themselves. In sum this work suggests that there are two ways in which care-givers' expressed emotion can affect health outcomes:

- by affecting compliance with treatment; and
- by inducing physiological change through stress.

Whilst some researchers in this field have examined the direction of causality (e.g. Gilhooly and Whittick 1989) in the main the focus has been on how informal carers' expressed emotion can affect the health and well-being of the care-recipient.

Health research more generally has focused on emotion and care-giver 'burden' a complex construct referring to the adverse effects of care-giving on the lives of informal carers. Though informal carers can derive considerable satisfaction from their caring role, there is now significant evidence that caring can contribute to both physical and psychological ill-health (Hirst 2003). Studies, for example have explored such aspects as the increased risk of burnout, stress, depression, loss and distress (see for example, Almberg et al. 1997; Denihan et al. 1998; Burns 2000). Work in this vein has highlighted how high levels of care-giving in the home, combined with inadequate or inappropriate forms of formal and informal support can lead to a breakdown in both the physical and mental health of the informal care-giver and hence the increased likelihood of a transition in the place of care from the domestic home to a residential care setting.

There has, however, been some debate about the *extent* to which caring for affects the health of the informal carer, particularly their physical health (although some carers face particular risks of injury associated with heavy caring activities). But as the 2001 UK census data clearly revealed, the greater the amount of time spent on care-giving, the greater the likelihood that the informal carer's own health and mental well-being will suffer. Indeed, the UK census illustrated that whilst 28 percent of those informal carers caring for less than 20 hours per week felt their health had been affected by their caring role, this increased to over 70 percent for those caring for someone for 50 hours or more per week (see Table 9.1).

The three most significant factors affecting the physical and emotional health of carers include tiredness, disturbed sleep and general feelings of strain. These in turn are likely to exacerbate other symptoms of failing health. Work by Hirst (2005), Keeley and Clark (2002) and others point out that those who are caring for

Table 9.1   Health symptoms felt by informal carers in Great Britain by number of hours spent caring (2000-2001)

	Number of hours spent caring per week			
	Under 20	20-49	50 or more	Total*
Feels tired	12	34	52	20
Feels depressed	7	27	34	14
Loss of appetite	1	5	8	3
Disturbed sleep	7	24	47	14
General feeling of strain	14	35	40	20
Physical strain	3	10	24	7
Short tempered	11	29	36	17
Had to see own GP	2	8	17	4
Other	2	4	2	2
Health not affected	72	39	28	61

Note: * Includes those who did not estimate the number of hours.

Source: General Household Survey, Office for National Statistics. Reproduced with permission.

more than 20 hours a week over extended periods of time will be more susceptible to such health problems such as anxiety, depression and psychiatric illness. Those undertaking more intensive care-giving are also more likely to experience lower social functioning and have an increased susceptibility to physical illness.

Carer burnout and the detrimental health impacts on informal carers are two of the primary factors underpinning informal carers' decisions to stop caring – arguably they underpin growing state recognition of the need support informal carers (see Maher and Green 2002; Wanless 2006). Maintaining the social and emotional health of carers is thus important not only for ensuring the quality of the caring relationship, but for ensuring informal carers are both able and willing to continue caring and cope with the demands of care.

The propensity to poor health as a result of caring for, however, will vary dependent on a range of factors that includes not only the intensity and duration of care required, but also physical proximity and distance – with co-resident carers more susceptible to poor health as a result of their care-giving activities that than distance carers. This will also vary dependent on the characteristics of the informal carer him- or herself (for example, age, sex, relationship to care-recipient and so forth). These effects will also be compounded by local variations in the extent (or lack) of support and respite the informal carer is able to access (Arksey and Hirst 2005).

Placing the Social and Emotional Accounts of Care

While the health and psychological literatures have tended to focus on the physical and emotional health impacts of 'carer burden', sociological and geographical accounts have sought to explore interactive narratives of emotion. These differ significantly in that they bring together the analytically distinct, but inseparable, dimensions of care not just as physical and emotional labour but also as a social relation (Hochschild 1979; Thomas et al. 2002; Brown 2003). Rather than being seen as a discrete and individual attribute, the relational nature of care helps to make carers and care-recipients 'how they are who they are' (Brown 2003, 834). At a conceptual level, then, relational care aims to describe 'the complex of activities, *feelings* [author's emphasis] and obligations that are involved in the support of individuals, usually family members' (Twigg 2000, 161), its purpose being to maintain, continue and repair bodies, minds and the 'worlds' we live in order to maximise an individual's ability to continue to live in it as well as possible (Tronto 1994).

Geographical work on emotions is a fairly new but emerging area (for an overview see Davidson and Milligan 2004; Davidson et al. 2005) and that which focuses specifically on the lives of older people and informal carers is very limited. Perhaps of most interest is Rowles's (1978) early geographical work on older people's experience of place. Whilst emotion was not the primary focus of his work, he nevertheless drew attention to older people's emotional attachment to places and the feelings they invoke. More specifically, he noted that 'Feelings about place may reflect sentiments ranging from dread to elation. Often they are amorphous, multifarious, or inchoate. On an intuitive level it is clear that feelings associated with place are an integral component of the participant's geographical experience' (1978, 174).

For Rowles (1978), emotional attachment to places can be encapsulated within a threefold typology and characterised either as:

- Immediate: highly situation specific and relevant for only a short duration.
- Temporary: of rather longer duration and often repetitive in character.
- Permanent: where there is stability in a deeply ingrained emotional identification attached to place.

Emotional attachment to places can also be classified as either personal, stemming directly from an individual's unique experience; or shared, involving the mediation of others in sustaining an intersubjectively experienced sense of place.

Hence, emotions attached to places may vary according to particular contexts that can vary across time and space. Homes and care homes that seem supportive and friendly for some may invoke anxiety and distress amongst other; a place that may seem safe and homelike may change with time and frailty to become isolating and clinical. Hence places in which care takes place can foster seemingly contradictory emotions.

By and large, however, the various works referred to above have focused on emotion and informal care as a private activity – one that takes place within the home. Limited attention has been paid to the interrelationship between emotion, care and place across the broader landscape of care. Further, as previous chapters have illustrated, there are questions about what we mean by 'home' and hence how emotion and emotional attachment to place and home is played out within the care-giving relationship. This becomes particularly significant when we begin to think about the breakdown of informal care for older people and subsequent care transitions from domestic to residential settings. The affective performance of informal care does not stop once the site of care shifts to the semi-public space of the residential care home. While the physical nature of caring might change, the affective elements of that care remain – and may even increase. This all points to the need for a deeper understanding of how place is integral to the socio-emotional relationships of care.

Placing emotions within the informal care-giving experience requires us to consider two distinct aspects of caring. Firstly, the embodied emotional experience of informal care-giving that involves the informal carer's (inner) felt response to care-giving and how it impacts on their own health and well-being. Secondly, the affective, or emotional, entity of informal care work that involves an understanding of how the informal carer interprets and responds to the needs of the care-recipient. This may involve working to control the outward expression of his or her own feelings and performing actions that may be at odds with the inner state.

Emotion, gender and care

It is perhaps worth noting here that while much of the literature around emotion and care focuses on the gendered (female) nature of care – one that is explicitly seen to be located within the domestic home (Graham 1991), to date, there appears little consensus about how men experience the affective entity of caring. Gollins (2004) however, suggests that men approach informal caring in a very different way to women, in that they see it as an activity that is unrelated to their identity. Indeed, he maintains that while male carers speak easily about the emotions they feel as husbands or partners, they specifically avoid expressing the affective aspects of caring. So while men may be happy to express those emotions deemed as 'natural' within a marital or partner relationship – such as love, desire, duty and commitment – they rarely discuss those other feelings of loneliness, sadness, bereavement or need that form a significant part of the care-giving relationship. Milligan's (2005) work on the experience of informal care-giving in New Zealand, however, indicates that, at least amongst spousal carers, men's experience of the affective entity of caring may not differ significantly from that of women. Baker (1992) and others, also maintain that amongst male spouse carers there is evidence of a high level of emotional investment in the caring relationship. This he ascribes to a sense of reciprocity – where men view the giving of care as a return investment for familial care given by women in the earlier years of their marital relationship.

Based on the concept of 'reclaimed powers' (Gutman 1987 in Baker 1992), Baker argues that men become more family oriented and nurturant in the post-parental years – a view that runs contrary to traditional gender role theory that ascribes the greatest emotional investment in the care-giving role to women. Whilst research is this field remains limited, the above cited work certainly gives at least some credence to these claims and as such it warrants closer examination.

The Socio-emotional Costs of Caring

Whatever the gendered nature of affective care relationships, the emotional costs to caring are also socially manifest. Wanless (2006) and others have pointed to some of the considerable indirect costs of caring – such as social exclusion and the erosion of personal relationships. The demands of care-giving within the home and the nature of some illnesses, for example, can lead to a considerable reduction in former friendship networks. The extent to which supportive family and friendship networks exist or decline will of course vary, but many carers point to a widespread lack of understanding and an 'inability to cope' with changes in the care-recipients' health or abilities amongst those they formerly looked upon as friends. As a result, these individuals can either stop visiting altogether or do so only infrequently. This is particularly prevalent where there is a decline in cognitive abilities – as in dementia – and hence a decline in the care-recipient's ability to continue a sustained conversation. As one carer put it, 'I don't know if anyone else finds this, but when someone's diagnosed with Alzheimer's, you would wonder if they've been diagnosed with *leprosy* instead because nobody comes near!' (Milligan 2006, 324). As a consequence both the care-recipient and informal carer can find themselves experiencing increased social isolation.

This can be compounded by the changing relationships that can occur between the informal carer and care-recipient. Wiles (2003), for example, points to the way in which frail older people not only have to struggle with the effects of increasing illness or disability, but that their relationship with the informal carer often becomes lost, or qualitatively changes. So on the one hand both carer and care-recipient can find themselves being pushed into sharing increasingly intimate personal care relationships that can act to change the spousal or parental relationship. At the same time, they can experience a reduction in other social activities that they might formerly have shared – such as hobbies, travel, or social outings with friends. As previously noted, the sheer effort involved in preparing for an outing, the challenges of attempting to negotiate transport systems or other inaccessible social amenities can severely curtail social engagement outside the home. While these constructs relate the social engagement of both informal carer and care-recipient together, even where the informal carer has access to some form of non-residential respite (for example, a few hours of in-home respite or daycare), without some degree of flexibility, the time constraints of this respite can place geographical limits on the informal carers movements.

Emotion and the Relationships of Care

In this section I discuss the importance of considering how emotion is experienced and performed in formal care settings. Consider these three separate accounts of care within institutional settings drawn from interviews with informal carers of older people:

> ... he'd become incontinent ... and oooh, how humiliating, he's sitting with this green gown on and the catheter hanging down out on the floor – it was just – he had no rug over him and no pyjama coat – it was just – ooh, I just felt so *humiliated* for him when I went in.

> They were dreadful. One of them said to 'Bert' – he was one of the ones who say 'yeay or nay' to whether he was going to stay or go [a doctor] – 'now if you don't get out of bed today I'm going to take a big stick to you'! ... it's just the *strange* way they treated him, like he wasn't a human being or somebody with a bit of intelligence, he was just 'that thing in the bed'.

> Staff used to try and feed two, three, sometimes four people at one time, and a lot of these people couldn't eat fast – and it was *shovelled* in! And of course some of them couldn't eat it that quick and I found that very distressing.

Whilst these may sound like care episodes that took place some 30 or 40 years ago, in fact these are drawn from narrative accounts written by informal carers less than five years ago. Clearly there is an element of 'shock value' to these accounts, but more critically they highlight not only a sidelining of the care-recipient's (and carer's) emotional distress, but that their care has been objectified as a task to be fulfilled by those involved in the delivery of formal care. Objectification of the older person can render unnecessary the requirement to take account of his or her feelings of dignity and self-worth, but it also acts to strip both the care professional and their elderly patient of their status as thinking, feeling and emotional individuals.

Hochschild (1993) maintained that such responses by formal carers can be been seen to represent a defence mechanism against the potentially damaging effects of emotional labour. By this it is meant that formal carers actively seek to depersonalise their care work. By not engaging emotionally with older care-recipients the care worker is able to protect him or herself from any sense of loss or emotional upheaval that may arise following the care-recipient's transition to residential care or death. Lee-Treweek (1996) however, maintained that not only do formal carers engage with emotion in relation to care-recipients, but they can use that emotional connection as a means of managing their care work. In her study of care homes, for example, she illustrated how emotional labour is extended to include the emotional manipulation deployed by formal care assistants in care home settings as a means of maximising their ability to manage their workload. She does, however, distinguish between emotional labour and emotional work. The former is viewed as non-autonomous, and at the behest of the organisation, while the latter is seen as being under the personal control of the worker. Emotional labour is thus seen to represent a mechanism through which order can be maintained in formal care settings. By employing elements of both nurture and control, care workers are able to develop techniques, autonomously from the official care regime, that enable them to exert control over the conditions of their labour. Through the giving and withholding of personal warmth and affection, for example, the (predominantly female) formal carers deploy what Twigg (2000, 163) refers to as a 'nurturant power' that facilitates their ability to control the behaviour of the care-recipient. Residents who resist this emotional manipulation – particularly the infantilisation that frequently forms a part of it – can find themselves labelled 'difficult', hence less worthy of affection.

It is also important to point out, however, that formal carers do not only engage with emotional work as a tactic for maintaining order and control. Despite the suggestion that they deliberately withhold or manipulate emotion for their own purposes, many develop strong emotional attachment to care-recipients within both domiciliary and residential within care settings. However, given that formal care support is generally targeted at some of the frailest older people in society, such emotional attachment can be a double-edged sword. On the one hand it can bring feelings of satisfaction and immense reward to what is often a frustrating, poorly paid and difficult job, on the other, it exposes formal carers to feelings of sadness, bereavement and loss – either through the transfer of the care-recipient to another home, or wing of the same home, where more intensive care can be offered, or through eventual death (Thomas et al. 2002; Fleming and Taylor 2007; Nordentoft 2008).

What these accounts illustrate is the relational nature of care, that is, how emotions are composed of and defined as a set of interrelated reactions, instrumental responses and subjective experiences that are determined by social rather than biological processes (Averill 1996). As Hepworth (1998) pointed out, these are essentially learned ways of responding to social situations, and as such, are open to transformation over time and place. Hence, the affective experience of receiving care and support cannot be understood as an individual socially-isolated phenomena; spouses, partners, other family members and care professionals all actively participate in shaping that particular physical and affective landscape of the care. As Thomas et al. (2002) put it, 'the practical and emotional involvement of these socially significant others in the patient's journey through illness affects those companion's own lives, sometimes in profound ways' (530). Importantly, *where* that experience occurs will also have a significant effect in shaping those experiences. This can be particularly so for those who are actively involved in care-giving.

Davis (1979) maintained that the attribution of distinctive emotions to the care-giving experience can be interpreted as a socially determined mechanism for maintaining boundaries between different groups of individuals, making it easier to make sense of the tensions between public perceptions and subjective experiences.

This distinction between public perceptions and subjective experiences also brings to the fore the distinction between the way emotion is expressed through the socially prescribed (outer) self – that is shaped by how we are viewed and valued in society, and the subjective (inner) sense of self (Hepworth 1998). This interplay between an individual's subjective experience and the outer world is linked to cultural prescriptions of what it is to be 'old' and requiring care – and the appropriate actions and emotions attached to the experience of giving and receiving care in that particular socio-cultural context.

Subjective emotions are, therefore, viewed in the social world in terms of culturally determined reference points and modes of expression. The duality highlights the tension between on the one hand, socially prescribed emotions attached to an individual's place in society and on the other, his or her private subjective feelings. The outward manifestation of the tension between the inner and outer self has been described as 'a mask' (Biggs 2004) – one that allows for the separation of inner feelings from their public and external expression (that presented to the world at large). This provides space for a cultural elaboration of the distinction of those emotions viewed as 'morally worthy' and hence acceptable to air publicly and those labelled as somehow 'disreputable' – to be squashed or at best hidden from the public gaze. Recognition of publicly 'acceptable' and 'unacceptable' emotions is reflected in the following comment from an informal care-giver:

> ... most people who work in an office or a shop or whatever, meet together and they can say, 'the kids are driving me mad' or 'the husband's driving me barmy, I think I'll go home and kick the cat!' or something like that. Nobody bat's an eye, its 'oh, ha ha, very funny!' but you try saying about someone who is disabled or got Alzheimer's, 'mum's driving me mad, I'm gonna' kill her, she's driving me round the bend!' – and its, 'Ooh, that's a terrible thing to say' – you should be a loving and caring person. And so you are, but you're only human ...

Whilst this carer clearly acknowledges and accepts that her emotions of anger and frustration are an inevitable part of the human care-giving experience, she also recognises that the cultural dictates of the society in which she is embedded require that she present herself as a loving, caring person – expressing only those emotions viewed as 'morally worthy'; to do otherwise would somehow brand her as 'morally disreputable'. The end result is a public expression of an 'idealised' self (the external mask) where social respect is wrapped up in an emotional labour that activates cultural images of 'respectable emotions' attached to specific groups of people in our society. The social performance of 'respectable emotions' thus requires the informal carer to engage in a self-conscious distancing from his or her subjective feelings. These social mores seem to govern how people try, or try not, to feel in ways that are seen as appropriate to the situation, and as such point to a social ordering of emotive experience (Hochschild 1979).

The comment above is linked specifically to fears that the expression of any emotion that could be interpreted as socially or morally unacceptable may result in societal action that would remove their right to care *for*. This same carer goes on to say:

> ... if a social worker or somebody goes in and asks the carer if you are coping, or 'how are you today?' you can BET your bottom dollar the answer's going to be 'fine' because they're not going to say to someone they hardly know, 'Oh, terrible'. There's that fear that mum, dad, in-law or whatever is going to be taken away.

Such complexities raise important questions concerning the precise nature of emotions associated with the care-giving experience and the role of place within this experience.

Emotion and the Care Transition

Previous chapters have drawn attention to the way in which the home constitutes an emotionally dense setting linked variously to feelings of identity, ontological security, a sense of belonging and for some, a sense of fear. As the need for care support increases it can also bring with it a sense of dislocation as the home takes on attributes of the clinical setting. For some, increased frailty can result in informal carer breakdown and transition to residential care as they are no longer able to cope with the increasingly heavy demands of caring. While the transition often brings mental and physical relief from the stress, anxiety and sheer physicality of care in the home; in almost all cases, this is tempered by what can be almost overwhelming feelings of guilt, failure, betrayal, loss, helplessness and anxiety. The range of feelings informal carers can experience during the transition of care and the subsequent impact it can have on their own mental well-being is summed up by one spouse carer below:

> The night before she went [into residential care] she had dozed on our settee and had leaned over to one side. I sat beside her and told her I was sorry I couldn't look after her any more, I helped her up, took her to the toilet, got her ready for bed and helped her settle. I went and sat in the lounge and cried and cried. I felt distraught, that I had betrayed her, I had let her down, I felt guilty, I was devastated and felt desolate ... I had an intense period of grief for a few months and although I am not a depressive person I had some of the physical symptoms (Milligan 2006, 323).

In this particular instance, the informal carer's feelings of guilt and failure were compounded by having to weigh up the advantages of ensuring his wife was able to enter the care home of choice and the recognition that this would mean her entering the care home some two months earlier than he had anticipated because a place had become available. Though two months would not seem an overly long time, it meant he was unable to share a significant event – a last Christmas – at home with his wife. His feelings of guilt were compounded by a belief that in transferring his wife's care to a residential setting he has somehow failed to maintain the reciprocal care to which he felt she was entitled. He goes on to remark, 'I felt I had no option but to take it. I felt awful, I didn't want her to go, she loved me and had cared for me all those years and I loved her and wanted to do the same for her but just couldn't' (Milligan 2005, 2112).

The stress and emotional angst of finding an acceptable care home should not be underestimated. But even where this has been achieved, further distress for both care-recipient and informal carers can occur where the needs of the care-recipient increase following initial admission, leading the care home to request that an alternative care placement is sought.

For spouse or partner carers, these transitions in the locus of care can also manifest in a significant change in their own experience of home. Despite knowing that the care-recipient is still alive, the informal carer has to adjust to living alone after what may have been many years of co-habitation. The dilemma for the informal carer is one of knowing that the care-recipient will be unlikely to ever return to the domestic home for more than an occasional visit. Yet they often feel unable to remove or pack away artefacts and objects belonging to the care-recipient. As one spouse carer commented:

> The hardest thing is the continual grief. I only need to open a cupboard and there may be something of Steven's and I feel devastated. I cannot dispose of all of his possessions because although he will never use them again, he is still alive (Milligan 2005, 2112).

Even given declining health and, in some cases the ability to communicate, carers can still feel a keen sense of the loss of companionship. As one carer put it, '... you don't need to talk a lot [it's] just having someone there.' The physical loss is also manifest in an inner sense of loss, a feeling that 'there's a sort of hollow where there's no-one there now'. Even where an informal carer has been non-resident, for example an adult child carer or sibling, the task of dismantling the care-recipient's home (which may also have been the childhood home) and disposing of possessions gathered over the course of a lifetime, can be highly distressing. For some informal carers, this period of transition is akin to the experience of bereavement but, as suggested above, without the finality and closure of death.

As noted on Chapter 8, not only can both the carer and care-recipient experience considerable stress and anxiety as care is relocated from the home to residential care settings, but this anxiety does not necessarily lessen over time. As the health and mobility of the older care-recipient declines this can result in shifts across the spectrum of care from relatively low-level supported accommodation in community settings and 'extra care' to residential care homes. For both informal carer and care-recipient these moves are indicators of a further decline in the care-recipient's health abilities and the need to readjust to the new care setting. As one informal carer put it, 'I don't know, there's a different atmosphere in [care home], there's atmosphere of almost dreariness and death ... and the other ones [residents] that are around him make noises and what have you.' These examples all point to the importance of understanding place and emotion in the care of older people.

The discussion so far assumes the existence of a close and loving relationship between the informal carer and care-recipient. This, of course, is not always the case. Where marital, partner or familial discord exists, informal care may be provided out of a sense of duty rather than a sense of love and reciprocity. The opportunity to retreat from caring may thus be experienced as a sense of relief and elation rather than guilt and regret. Though this is the exception rather than the rule, one elderly spouse carer in an unhappy relationship noted: 'he was in for respite care, they didn't let him come home, he was in and was staying in and I was free. ... Would I bring him home? Not on your Nellie!' For most informal carers, however, the shift in the place of care is marked by a sense of 'displacement' characterised by such emotions as sadness, guilt, a sense of failure and anxieties about potential changes in the relationship between themselves and the care-recipient.

The Place of Carer Support and Coping Mechanisms

How do informal carers deal with the social and emotional upheavals of caring? The increasing social isolation and breakdown of previously existing social networks means informal carers have to look elsewhere for support. Clearly formal care services are critical in facilitating this, but one important source of support that is frequently referred to by informal carers is that of other informal carers. Carer support groups or contact with other informal carers made whilst visiting day-centres or institutional settings can facilitate the development of new friendships. Such friendships can spill over into the everyday lives of informal carers resulting in social contact over coffee, lunches and other social outings. In some case informal carers become so enmeshed in carer support that they themselves begin to take on support and counselling roles for other, 'newer', informal carers. As one ex-carer noted: 'I have people ring me, I have them come and see me, I keep in touch with them if I don't hear from them and I ask them what they've been doing, and what are they going to do and what shall we do? And just without being too invasive, I try and move them along a little bit, you know?' Whilst carer support groups do not suit all, they can provide a valuable mechanism for developing new friendship networks. They can also be an important repository and source of shared knowledge and information about available care support, residential care homes, processes and practices. Sharing experiences with other carers can also help informal carers to recognise that the feelings of frustration, anger, guilt, betrayal, grief and so forth that they may experience during their caring lives are not unique. The knowledge that other informal carers experience similar feelings can be important in helping them to terms with their own experiences. In discussing a carer support group one individual noted: 'I've met so many people – all more or less in the same boat – and we all support one another. You meet people and you listen and talk and you hear stories and you have a laugh and have a few tears and listen to different things … and you pick up all sorts of things there about the various homes – so that's *good* support.'

Voluntary organisations, whose remit includes carers and older people, can also provide a valuable source of support, advice and information about the processes and practicalities surrounding the availability of formal care support services in the home and community, benefits, the process of transition to residential care and any subsequent transitions.

Whilst each experience will vary, it also appears that where decisions are shared, this can help to reduce some of the most insidious (negative) feelings attached to informal caring. Being able to share the responsibility with close family members, for example, can be an important factor in helping the primary carer to cope with undertaking difficult or major decisions – such as entry to residential care. As one informal carer commented: 'I couldn't cope day and night … so I had to sign papers, which I wouldn't, but my son did for me. I couldn't just take that responsibility and put him away. So with my consent he signed.' However, not all informal carers have the support of close family. Indeed, in some instances there can be a clear lack of understanding between family members of the extent of care support that is actually required on a daily basis and the inability of the informal carer to continue providing that support. In such cases, informal carers can feel both an additional burden of guilt that they are being seen not to 'care enough' by those family member who are removed from the daily task of caring, and a sense of anger towards these other family member at their lack of recognition of the extent of care work that is required to sustain the care-recipient at home. As one carer noted:

> I found I got very angry at my brothers at this time because they left me to do the 'donkey work' … [but] I'm happy to have done what I did for my parents. I always thought of it as payback time for their caring for me. I never resented them – just my brothers.

In conjunction with these experiences, informal carers have also pointed to the important role that health professionals can play in helping to 'legitimise' their decision to stop caring *for* within the domestic home. By voicing their opinion or judgement that a care-recipient's health has deteriorated beyond the ability of the informal carer to manage at home, health professionals can help the informal carer to justify their decision to other family members. Such action has the additional effect of helping to relieve the informal carer's feelings of guilt and failure. Whilst some carers may feel such action by health professionals can push them into a decision they would prefer not to have made, many more feel that by adopting this approach, health professionals are in fact helping to relieve them of some of the burden of decision-making.

Finally, following the move to residential care, spouse carers not only have to come to terms with lone-dwelling and the loneliness that accompanies the loss of their lifetime companion, but they also have to find ways of filling the gaps left in their lives following the 'hand-over' of many of the caring tasks they formerly performed to paid care staff within a care home setting. For some, this can initially involve catching up on household and general maintenance tasks that had been neglected during their time spent caring in the home. For others this can involve a conscious decision to immerse themselves in new activities and tasks such as walking, gardening, golfing etc. As one ex-carer noted:

> For years I baked, until I had a deep freeze *full* of food! I'd make some muffins, made some biscuits. I didn't want them, didn't need them! I gave them away, eventually. But *do* something, or think to yourself, now every day something good has to happen in your life.

Concluding Comments

As a set of interrelated reactions, instrumental responses and subjective experiences emotions can clearly be seen to be determined by social rather than biological processes. Little attention has been given to the socio-spatial mediation of the emotional aspects of care thus far, yet it is evident within this chapter that this is an important and integral part of the care-giving experience. Emotions not only affect feelings *about* places, but where that care takes place impacts on the emotional lives of those involved in the care-giving experience. Earlier literatures tended to view the emotional work of caring as something that was not only highly gendered, but also largely site specific – being located almost exclusively within the private space of the home. However, it is clear that the affective nature of care stretches across space, from the domestic home to the residential care home and well beyond. Indeed, as this chapter has illustrated, the affective nature of care is particularly evident when *transitions* in the place of care take place, and can trigger a whole series of embodied emotional responses. The chapter also suggest that assumptions about the gendered nature of affective care needs to be revisited – at least in relation to care and older people. Finally, though not specifically discussed in this chapter, the unbounded nature of affect in relation to care means it is not limited by proximity or distance. Understanding the socio-emotional impacts of caring and their potential impact on the lives of both carers and care-recipients is thus important if we are to have a greater understanding of the landscape of care.

Chapter 10

Reconfiguring the Landscape of Care: Porosity, Integration and Extitution

At the outset this book pointed to the way in which the shift to 'ageing in place' has increased the complexity of the care-giving relationship such that it can no longer be understood through a single, situated narrative. As previous chapters have revealed, these narratives are manifest through a series of interwoven stories that emerge at different times as a result of differing sets of circumstances between formal and informal caregivers and older recipients of care. These narratives do not occur in discrete spaces, but stretch across and beyond the domestic home, to include both community and institutional settings. In this final chapter, I consider whether these interwoven stories are giving rise to an increased porosity not just between the worlds of formal and informal care but also across the spaces within which that care occurs. This in turn gives rise to discussion about those factors that contribute to the emergence of inclusionary and exclusionary spaces of care. Finally, the chapter returns to ideas of institution and extitution to consider how shifts in the contemporary landscape of care for older people can be conceptualised within these debates.

Places, Practices and Porosity

Discourse on the blurring of the boundaries between formal and informal care has been evident for some while. Commentators such as Ungerson (1990) and Thomas (1993) for example, pointed out over a decade ago that the conceptual divide between formal and informal care posed a false dichotomy in its assumption that the nature of the relationships that prevail within these two spheres are totally different. Indeed, Thomas (1993) claimed that the differing boundaries drawn around what constitutes care had the effect of excluding or including particular sets of social relationships. This suggests that viewing the worlds of formal and informal care as discrete entities is likely to cloud our understanding of the complex relationships that exist between them. Moreover, it overlooks the way in which the work of care can transcend the bifurcation between public and private, requiring us to both rethink the divide between formal and informal care-giving.

Care and support for older people consists of a complex arrangement of formal and informal provision that may be delivered within either or both domestic and non-domestic settings. So for example, those requiring care may receive it within the domestic home from a co-resident or ex-resident informal carer, a paid care worker, health professional or volunteer; they may receive day care in community settings; 'distance care' from both formal and informal carers through the implementation and utilisation of new care technologies; and respite care within the home of a paid care worker, or more commonly, within a residential care home. These care arrangements can occur across a range of different time-frames that include daily, intermittent, or continuous care support. Finally, as levels of frailty increase, care arrangements for the older person may be located entirely within a residential care setting. While these residential settings may be publicly, privately or voluntary sector owned, care work is predominantly undertaken by paid care and nursing staff. Nevertheless, as noted in Chapter 8, informal carers can still undertake a significant care-giving role within these institutional settings. In thinking about the complexity of the care relationship then, it is important to consider not just the relationship between formal and informal care, but how that relationship is manifest through the spaces in which that care takes place.

The continued development of care policies designed to support ageing in place over the last decade or so has meant that the relationships and practices of care – and where they take place – have become increasing porous. Within the domestic home, for example, informal carers are increasingly being viewed as co-workers who perform a range of tasks formerly seen as the exclusive province of trained formal care-givers (such as monitoring, medicating, injecting, catheterisation, fistula care, the changing of dressings and so forth). Indeed, Ward-Griffin and Marshall's (2003) work in the Canadian context found that care work in the home was continuously being transferred from nursing staff to informal care-givers with the expectation that over time, informal carers would assume increasing and more complex responsibilities. Here, formal care staff were seen to intentionally lessen the distinction between professional and informal caring in order to transfer certain types of caring labour to the informal carer. In other words, the boundary between formal and informal care-giving was becoming increasingly porous. Importantly, however, this study also revealed that nursing staff employed an element of persuasion – and in some cases coercion – to get informal carers to take on these tasks even when they were clearly reluctant to do so. As they note, 'some of them [informal carers] did not cross this boundary willingly, complaining or feeling frightened, overwhelmed, or angry that they were expected to provide care requiring "higher" care usually associated with nursing' (Ward-Griffin and Marshall 2003, 203).

Such work points to the way in which, within the domestic home at least, informal carers, as co-workers, are increasingly substituting for or replacing care originally provided by care professionals. This shift, furthermore, is seen as an intentional move on the part of the formal care services. Critically, however, once the care-recipient makes the transition to residential care, these care tasks once again become appropriated and performed by formal care workers. This highlights the way in which care tasks that are being actively de-medicalised within domestic space are being consciously re-medicalised within residential care settings in order to fit within the organisational construction of institutional care. *Where care takes place, then, actively contributes to the way in which that care is being constructed and reconstructed.* Though informal carers can undertake a substantial care-giving role within residential settings, re-medicalisation is effectively excluding them from the hands-on tasks of care-giving that they may have performed (and indeed been encouraged to perform) within domestic settings. Nevertheless, informal carers can still be seen to exert agency through the negotiation of new caring identities for themselves and/or the adoption of covert tactics in order to continue their care-giving activities. In this way, care-giving can be seen as being continually negotiated and renegotiated across the boundaries and spaces of care in ways that challenge conventional linkage models. Despite evidence of increasing porosity, however, the way in which care tasks are being both de-medicalised and re-medicalised to suit the conditions and requirements of health and social care professionals does highlight the ways in which power is differentially exerted in the renegotiation of these boundaries as well as where and how that power is performed.

The melding of the worlds of formal and informal care is also evident through the physical manifestation of the site of care. On the one hand this is marked through the intrusion of the medical paraphernalia of care within the domestic home and on the other – albeit to a lesser extent – through the intrusion of the paraphernalia of home into the residential care setting. Whilst the extent to which these intrusions occur may be asymmetrical, they are nevertheless indicative of an increased porosity not just between the worlds of formal and informal care but also between home and Home. The landscape of care for older people then, includes not just the interpersonal relationships that exist between carer and care-recipient – including here both familial and non-familial care-giving across the range of care settings, but also the concepts of home and Home as manifest within the caring relationship.

One outcome of the increasing complexity in the spatial dimensions of care for older people has been an incursion not only of those values, rationalities and temporal structures that make up the formal world of service provision into the private world of the home, but also the technologies and practices deemed necessary to perform that care (Twigg 2000). What this book highlights, however, is that whilst work by Twigg and others has been important in drawing attention to the increased porosity of the boundaries between formal and informal care in the private domain of the home and some of the tensions this gives rise to; this is not just a feature of home. There is evidence that this is also occurring within and across institutional and semi-public spaces of care. Indeed, while the porosity of care-related activities is evident across the spheres of home and institution, it is also evident within the intermediate or virtual spaces in-between (for example, informal care support within voluntary day-centres and informal carers' engagement with web-based care technologies and so forth). Hence within at least some of these settings we are also seeing a growing incursion of the values, rationalities and temporal structures that make up the *informal* world of care and support into community and institutional worlds of care. Shifts in care and support for frail older people can, thus, be seen to be encroaching upon and re-ordering those spatial divisions that have traditionally separated both the public and private and the professional and informal worlds of care.

The increasingly porous boundaries between who performs the tasks of care and the sites in which that care takes place points to the need for a more nuanced understanding of how and where contemporary care-giving is manifest. These newly emerging and more complex forms of care-giving not only transgress the boundaries between formal and informal care, but also the public and private spaces within which that care is performed. What is of particular interest to geographers about these shifting conceptualisations of the caring relationship is that they employ explicitly spatial expressions of care that draw on those particularistic forms of care-giving that exist between both formal and informal carers and cared-for within the private space of the domestic home and care given within public and semi-public spaces.

Porosity and Integrative Care?

A critical question raised by the shift noted, above is whether porosity and the move toward a co-worker model of care is indicative of a more integrative model of care and what might this mean for older people as recipients of that care. The shift to the politics of neo-liberalism in the UK (and many other advanced capitalist states) is manifest in devolution of central control in favour of local governance. Partnership working and 'bottom up' systems of care management and social regulation have become the desired norm. Policy discourse increasingly encompasses the rhetoric of citizenship, social inclusion and participation with an emphasis on choice, partnership and trust that seeks to challenge notions of welfare dependency. Yet despite the rhetoric of choice and partnership, as already noted, the reality is that most older people and their informal carers have a limited range of options and choices in relation to the services they receive, the extent or time of its delivery, or who delivers it. Indeed, for Powell and Biggs (2000), rather than initiating inclusion and choice, the current management of care represents a technology of regulation and control. In assessing, probing and inspecting a distinct population group (older people), they argue that care professionals have become interventionists in societal relations and in the management of their social relations. By playing a key role in ensuring individuals needs are regularly reviewed and resources effectively managed, care professionals probe and make normalising judgements about older people and their informal carers through discourses that define them variously as 'users', 'clients', 'patients', co-workers and so forth. In the UK, access to formal and contracted voluntary care services (unless privately purchased) is dependent on being assessed as meeting a threshold for care defined by the local state. Assessment is, thus, seen as a central technique that renders the individual an object of power/knowledge (Foucault 1977). The social practices of these older people and their informal carers are judged by 'expert' care professionals through a process of 'panoptic technology' in which older people are both the subjects and objects of the 'medical gaze' (Powell and Biggs 2000, 6). So despite the fact that policies around ageing in place are being defined in terms of empowerment, choice and so forth, Powell and Biggs argue that it is, in fact, part of a 'disciplinary strategy' that extends control over the minutiae of everyday life and conduct (11). While the space in which this 'gaze' operates can shift from the domestic home to the institution (including the residential care home), these spaces become sites for intense surveillance as well as the attainment of knowledge in relation to the carer and care-recipient.

From this perspective, porosity can be interpreted in terms of a deepening penetration of the state into the everyday lives of its older citizens and their informal care-givers (Wolch 1989). There is undoubtedly some value in this approach to understanding the shifting landscape of care for older people as well as debates around inclusion and regulation in relation to care. But while this approach requires us to take a more critical evaluation of where power lies in the relationships of care, it does presuppose a care relationship that is almost uni-directional. That is, power is understood to lie almost wholly with care professionals and policy makers. Yet as discussed within various chapters of this book, we should not forget that frail older people and their informal carers are not simply passive recipients of care, they can – and indeed do – exert both agency and reciprocity in the care relationship. Debates around inclusion versus regulation thus need to take a more nuanced approach to the unpacking of issues of power within the care relationship.

Inclusionary or Exclusionary Landscapes of Care?

While care legislation indicates that informal carers should be viewed as *partners* in the care of their spouse or relative, as suggested above, it can often be a difficult challenge for a family member to realise a sense of partnership where care 'professionals' play a critical role in assessing care and support needs. This highlights the need for a better understanding of how both inclusionary and exclusionary landscapes of care are constructed and how we might facilitate the better integration of formal and informal care. This section thus draws on case material to illustrate some of the key issues that can facilitate or inhibit the development of more inclusionary landscapes of care.

Laying the foundation for an inclusionary care environment

As noted in Chapter 8, relocating a spouse or close family member to a residential care setting can be a traumatic event for an informal carer, hence how they, and the care-recipient, are introduced to the new care environment is important for the well-being of both. Effectively achieved, this can lay down a foundation for the development of a good working relationship between the informal carer and formal care staff. Being introduced to the care environment, having the care home processes carefully explained at an early stage, along with a two-way discussion of how the care of the spouse or close family member is to be jointly managed by both the care home and the informal carer can ease concerns about their role in this new place of care. Paid care staff also need to be cognisant of, and equipped to deal with, the range of emotions – from sadness, confusion, a sense of loss, guilt, anger and betrayal and so forth – that informal carers as well as care-recipients are likely to be experiencing at this point in the transition of care.

Yet it is important to recognise that informal carers, like care-recipients, are not a homogenous group, hence they will have differing views and expectations about care homes and what they consider acceptable care practices within them. Having the 'people skills' to work and deal with these differing views and expectations is critical to 'good care'. One such example relates to the damage to personal belongings and the loss of, or confusion over clothing (despite being labelled) within care home settings. For some, finding the care-recipient dressed in other people's clothing, can be extremely distressing and indicative of a loss of the care-recipient's personal identity within the care home setting. Others, however, accept that such incidents are symptomatic of the difficulties of group living in an institutional environment. Yet where informal carers address such issues with staff in a calm and pleasant manner, and where staff, in turn, treat their objections with respect, a satisfactory outcome for all is more likely to be achieved. As one carer commented:

> Occasionally someone else will be wearing something of mum's, a minor issue. I handle it by saying to the nurses how nice so-and-so looks in that blouse (or whatever) but I think it belongs to mum. They usually offer to take it off right away but I won't hear of that, it's not that important. The item is usually returned after washing (Milligan 2004, 56).

Further, while there can be little excuse for a care-recipient to consistently appear in clothing that belongs to other people, or which is damaged or dirty, if clear explanations about what constitutes appropriate clothing, and why it is helpful for such clothing to be provided are made to the informal carer prior to care home entry, it can relieve some of the tensions discussed above. For example, clothing deemed appropriate when the care-recipient was mobile, more flexible or continent and so forth, may become inappropriate when it is important to simplify the dressing and changing process. Similarly, the need to wash clothing at high temperatures due to soiling renders certain fabrics inappropriate or liable to damage in the laundering process. Unless adequately explained, however, what may seem self-evident to those working in institutional care settings can be both bewildering and upsetting to those with little experience of them. As one carer noted:

> Occasionally I think we feel that a lot of mum's clothing *sits* there, because it's not as convenient to put it on as some of the other things she's got, if you know

what I mean? Its not just the washing thing, it's the physical, because mum isn't able to be very co-ordinated – you know she can't move forward for you to get her jersey on, you know, this sort of thing? There are obviously certain items of clothing that are easier to put on (Milligan 2004, 56).

Where the informal carer has been heavily involved in care-giving prior to care home entry there is likely to be a greater tendency to empathise with formal care staff and make allowances for occasional mistakes and mishaps as they recognise and acknowledge the difficult task that paid carers undertake.

As noted earlier, the monitoring role that informal carers perform is critical in ensuring that the quality of physical care given is of an acceptable standard. While informal carers need to be aware and informed about what they can realistically expect from a care home, it is important that they remain alert to unacceptable care behaviours. Whilst in general care homes provide good physical care and most minor complaints are dealt with effectively by care staff or care home managers, it is important, that where poor practices *do* occur, informal carers are fully aware of what constitutes unacceptable practice and the avenues through which they can be effectively dealt with.

Constructing inclusive spaces of care

Developing inclusionary models of care requires consideration of how informal carers can be made to feel a valued part of the new care environment. Some informal carers undertake intensive and/or many years of caring at home prior to the transition to a residential setting, and as noted in Chapter 8, their identity becomes intimately bound up with their caring role. Yet informal carers can be made to feel awkward or sidelined undertaking care work they would formerly have undertaken within the domestic home. The shift from caring at home to care delivered within an institutional environment can thus leave the informal carer feeling 'out of place' and unsure of their new role within the care setting. Hence within these settings the balance of power subtly (and sometimes not so subtly) changes from the informal carer and care-recipient to formal care-givers. Given that the extent of informal care-giving in residential settings can be surprisingly high, it is important that care homes find ways to ensure informal carers feel empowered to undertake these roles in the new care setting should they desire. The extent to which care homes achieve this goal, however, varies. Some make considerable efforts to ensure informal carers feel comfortable caring for their spouse or close family member in the new care environment:

I'm treated like ... oh, pretty much like staff I should think! I mean they know me by name, and in fact they know me better than I know their names! Yes, it is important that you still feel that you're still caring for her somehow, and that's one of the reasons I still feel comfortable going.

> A few times mum would have an 'accident' and if there was no-one around I would begin the task of showering her until someone took over. They didn't mind and I was never made to feel like I was interfering (Milligan 2004, 58).

Others, however, can find it a difficult challenge to continue to care in an institutional environment, and find it difficult to undertake even some of the very simple care tasks they would formerly have undertaken at home.

Constructing inclusive spaces of care also extends to: i) how decisions are made with regard to the care of the resident and potential changes in that care; and ii) the extent to which care-recipients (where able) or informal carers have access to medical and care records relating to the care-recipient. Prior to care home entry the informal carer, together with the care-recipient will have taken responsibility for all major decisions regarding care and treatment – including the decision over transitions to supported accommodation or care home entry. These decisions of course are likely to have been undertaken in consultation with, and on the advice of, health and social care professionals, nevertheless, unless no informal carer is evident and the care-recipient is deemed unable to make decisions, final decision-making will lie largely outside the sphere of formal care.

On entry to the care home the balance of power in relation to decision-making also changes. While care-recipients and informal carers are still involved in this process, generally though regular review meetings, health and social care professionals take a far more influential role in decisions around care and treatment than would have been the case prior to entry. Involvement in the development of care plans for the care-recipient however, can vary. In my own research, while some informal carers were aware that these plans should exist, few actively participated in their ongoing development. For example, when asked about his involvement in the care plan, one spouse carer commented:

> Yeah, that's what I haven't *seen* yet. I presume they've got one, but whether it's written down or not, I don't know. It should be I think. Yes, that came up – I saw something about that about a month or so ago, about a care plan. In, er, a consumer magazine or something, and I meant to ask, but I forgot ever since (Milligan 2004, 60).

Examples of good care practice in care homes illustrate, however, how care plans can be drawn up as a jointly negotiated document between care staff, the informal carer and care-recipient (where possible). These are revisited at regular intervals depending on changes in the health status of the care-recipient. Some care homes also ensure daily reports made by care staff are left in an easily accessible location for the informal carers' inspection, should they require it.

It is important to note however, that in some cases the lack of knowledge about the existence of records and care plans arises because informal carers feel satisfied with the level of care and information given, and hence feel no real necessity to become involved in care planning. In other cases this appears to be a consequence of carers' preference to confront care staff directly where they feel some aspect of care is unsatisfactory or that the health of the care-recipient has changed. Whatever the reason, with few exceptions, informal carers point to a decline in any 'partnership' role between care staff and informal care-givers in the care home and its replacement by a process of 'advisory' informal discussions on the day-to-day care of the care-recipient. For some, this process works well, as one informal carer noted:

> Was I included in these decisions? At every step of the way, both in the caring and medically. I mentioned things I saw about mum's changes and they had always noticed them too. The nursing manager and I would stand [informally] and have a chat almost every week and I always felt I was 'kept in the know' (Milligan 2006, 328).

Others, however, feel this process effectively excludes them from the decision-making process, noting that they are only advised of the progress of the care-recipient if they actively seek out a formal care staff themselves. Others point out that they are almost wholly reliant on care staff to inform them of significant changes in the health or well-being of the care-recipient during times when they are unable to be physically present. Importantly, what are viewed as significant events for informal carers (such as an accident or downturn in the health of the care-recipient) can be viewed more prosaically by formal carers, resulting in their failure to contact and advise the informal carer of the event.

Communication

The above comment illustrates the way in which informal carers perceive inclusion to be linked to good communication, the perceived accessibility and approachability of the care staff and the extent to which they ensure the informal care-giver is kept informed of changes and events in the life of the care-recipient. Of course, this is a two-way process, and informal carers, themselves, need to ensure care staff are aware of the level of information they require. As suggested above, where care and communication is perceived to be good, most informal carers are satisfied to be contacted only where significant events occur. Regular meetings between formal and informal carers (and where possible, care-recipients) can contribute to a greater level of trust and understanding between them and engender a greater feeling of partnership in the care relationship. As such, it is a practice to be encouraged. Conversely, the failure of care homes to inform informal carers of events in the lives of the care-recipient that they perceived to be of significance can lead to a breakdown in trust. As one carer noted:

> Last October I was taken by surprise on one visit to Mum because I was met at the entrance by one of the staff, who said, 'We're so pleased to see you ... you're mother thinks you have died and has been crying all week.' I turned the corner

and another staff member said almost the same thing, and as I went into Mum's
room, a third staff member said, 'you're mother will be so pleased to see you'
and proceeded to repeat the same thing. Why oh why didn't somebody phone
me? (Milligan 2006, 328)

Even where good communication mechanisms are in place, it is important that
discussion is structured and delivered in ways that are clear and understandable to
both informal carer and (where appropriate) care-recipient. It is also important that
the atmosphere is conducive to a two-way exchange of information and ideas.

Yet good communication involves more than just the interchange between
informal carers and care staff. It also includes the extent to which information
related to the care needs of a care-recipient is conveyed *between* formal care staff.
Indeed informal carers have noted that good communication between formal carers
about the specific care needs of individual residents is critical to their well-being.
As one carer commented:

There was two ways that I thought she could have been sick, one that was that
she had had a seizure and had been sick when nobody was noticing, *or* that
she had been given gluten. So I went and saw them, and I said, look, I don't
know whether she's been given anything but there's a possibility somebody
has given her some flour. And that's what could cause her sickness, and I think
it's important that new people know – and there are always new people now
(Milligan 2006, 329).

Pay and working conditions within care homes are often poor, leading to high
levels of staff turnover (Eborall and Garmesan 2001); hence this can be particularly
important in relation to new staff. As the interview excerpt above suggests, where
they have not been fully apprised of the specific care needs of an individual this
can result in potentially negative health outcomes. Furthermore, the discontinuity
of formal carers can be confusing for both informal carers and care-recipients,
inhibiting the ability to build good and inclusive working relationships and
reducing understandings of how formal and informal carers might usefully work
together for the benefit of the cared-for.

Carers as experts

As suggested in Chapter 8, inclusionary care environments need to recognise and
acknowledge the expertise of informal carers within the network of care. Not
only are they now performing care tasks in the domestic home that were formerly
medicalised, but their unique relationship with the care-recipient means that
they are able to offer crucial insights into the social, cultural and biographical
background of the care-recipient that, acknowledged and usefully employed, can
contribute to an improved quality of care. This can be particularly important in
residential homes, where care is delivered in a setting devoid of the normal visual

clues available within the domestic setting. Where care-recipients are unable to communicate easily, the close attention informal carers pay to changes in their behaviour patterns or their ability to recognise specific signals of distress due to their intimate knowledge of the individual can be particularly important, and where acted upon can help to alleviate particular health conditions or discomfort. As one carer noted:

> When mother was getting her urine infections, she used to get very agitated, and I always knew what it was, and I would say to them, 'mother's got a urine infection' and really once again they would said to me, 'how do you know?' and I would try to explain to them how I knew – and they'd say that because she wears a napkin, its very difficult to take a urine sample, and it normally took about fours days – and finally when it started bleeding, they would realise that she really did have a urine infection and she would be put on antibiotics (Milligan 2006, 326).

Such insights are also important in helping paid care staff to understand how and why a care-recipient may respond to situations or activities in specific ways as well as what may help to stimulate the care-recipient. So while paid carers can offer expertise in the physical care of the individual, informal care-givers can offer crucial insights into the wider social, cultural and biographical background of the care-recipient that are essentially and 'unknown' to paid care staff and can be important in contributing to the overall quality of care.

In order to facilitate the delivery of more inclusionary care within care home settings, then, care home staff need to think critically about how they can effectively integrate informal carers into the care home setting in ways that will enhance the nature of care delivered to the care-recipient as well as contributing to the mental well-being of the informal carer. Positively utilised by formal staff in these care settings such expertise can contribute significantly to inclusion and the overall quality of care.

Revisiting the Spatial Arrangements of Care: Institution, Extitution and Deinstitutionalisation

In the course of this book I have touched on ideas of institutionalisation, deinstitutionalisation and extitution as they relate to care for older people in the home, community and care home settings. I set out from the premise that community care – and ageing in place in particular – with its mix of formal and informal care support was seen to represent a significant physical and ideological shift away from earlier models of care. These earlier models saw care for older people lying either with the family within domestic settings (unsupported by formal care services) or with the state within large single-site residential settings. Much of the literature on this topic (both contemporary and historical) has focused on the effects of this shift in the location of care by drawing on the concept of deinstitutionalisation. Indeed, within the discipline of geography itself, a whole strand of work on this theme emerged from the mid-1970s onwards (see for example, Wolpert and Wolpert 1976; Dear and Wolch 1987; Milligan 1999; Wolch and Philo 2000; Hall and Kearns 2001; Joseph et al. 2009). Whilst much of this work has focused around mental health, it has also included work on older, disabled and learning disabled people. At least some of this work explicitly recognises that what we have in effect seen is the rise of an 'institution without walls'. As Vitores (2002) points out, the assumption that moving care beyond the physical structures of the institution would somehow also manifest in the extinction of the institution – not just as a physical structure but as a set of embodied practices – was inherently flawed. Indeed, Powell and Biggs (2000) elucidation of the 'panoptic technologies' and 'disciplinary strategies' employed in the management, regulation and control of contemporary care for older people is redolent – at least in part – of Goffman's (1961) earlier discourse on the 'total institution'.

Goffman's work in the 1960s drew attention to certain characteristics of institutional environments that marked a clear differentiation between daily living in an institutional environment and the domestic home. The former, he maintained, was characterised by lack of choice and personal autonomy, homogeneity of daily living and a clear distance between the world of the paid care worker and the care-recipient – in particular the secondary status of the care-recipient. While Goffman considered home life to be incompatible with institutional life, his argument was based on the view that the home was steeped in affective relationships that were absent from the institutional environment. Hence, he argued, that while residents and care staff might be intimate, there was a tendency to retreat from such a relationship because the engendered feelings that arose were inimical to the undertaking of care work. As more recent work by commentators such as Lee-Treweek (1996) and Skilbeck and Payne (2003) amongst others have pointed out, however, care staff within institutional settings can – and indeed do – sometimes become engaged in quite close affective relationships with those they care for. Indeed, as noted in Chapter 2, emotional care and support can be a critical component of some forms of paid care work. Furthermore, as the earlier discussion on porosity illustrates, teasing out the boundaries between home and institution and between formal and informal care is now far more complex that the earlier work by Goffman infers.

Hence as Powell and Biggs' (2000) work suggests, rather than thinking of the institution as some physically bounded entity, we should perhaps be thinking in terms of the institutionalising *performance* of care and how that is manifest across a new landscape of care that stretches from the home and community-based settings to the care home and beyond. As chapters within this book have illustrated, for example, within the domestic home, as care needs increase, the institutionalising performance of care is reinforced as tensions emerge between home as a site of identity and belonging and home as a site of work that requires a reshaping of the home to make way for the demands of formal care work. The balance of power between formal care workers care-recipients and informal carers shifts as formal carers are increasingly able to penetrate and reshape the home. The end result is a physical manifestation of institutionalisation within the domestic home. Furthermore, without careful consideration and planning, the growth of new care technologies runs the risk of further exacerbating this institutionalising process. This raises critical questions about the point at which those features of home seen to facilitate ageing in place become subsumed beneath the institutionalising process.

This is, of course, a partial story. As earlier discussions in this book have also illustrated, where the balance of power and control lies in the care relationship is to some extent a feature of where care takes place and the extent of care and support required. So where initial care and support needs in the home are minimal, power and control can lie firmly in the hands of the older person and his or her informal carer. The institutionalising performance of care is thus likely to be largely absent. Even where informal carers and care-recipients are dependent on formal care support, their daily life will still be characterised by heterogeneity, personal autonomy and choice. Within this environment, then, the status of the care-professional is reduced, increasing only as care needs increase resulting in an increasing shift toward an institutionalising environment described above – whether that be within the home, community or a residential care setting.

Concluding Comments

In sum, the discussion above indicates that rather than seeing care in the community and ageing in place as processes of *deinstitutionalisation*, the construction of formal and informal care within this process means that what we are in fact seeing is the emergence of new forms of institutionalisation based on performance. This can manifest within and beyond the physical structure of the traditional institution. While the new institution can at times be visible within the home, it can also emerge within and across other community-based spaces of care as well as in residential care settings. The growth of monitoring and surveilling care technologies is adding a further layer of complexity to the new institutionalising process. This is also marked by an increasing porosity of the boundaries between public/private formal/informal care and where the balance of power lies across these boundaries. This process can be more accurately wrapped up in the concept of extitution.

Chapter 11
Concluding Commentary

This book set out to contribute to a new geographical analysis of care for frail older people – one that weaves together the narratives of care for older people across a landscape that stretches from the home to the community, residential care and beyond. This perspective has largely been unexplored by those researching care through historical, sociological, gerontological or social policy lenses. The theoretical and empirical analyses offered here reinforce how care for older people is laden with territoriality. As such, it can be seen as a spatial expression of how human action is bound up not just with the power relationships of care, but also with tensions, conflict, emotion and change. How these spaces are manifest then, can be seen in part, as the outcome of reaction to the territorial action of one party (for example formal or informal carers, or care-recipient) and in part as a response to attempts to modify the spaces and places in which that care occurs. Such shifts, I have argued, can be read and understood in part, through the extitutional arrangements of care, and in part through the concepts of anthropological and non-place and their relationship to notions of home and Home.

However, as the book has further demonstrated, the experiences of care and care-giving for older people are also culturally and politically constructed. These constructs have implications for the extent to which the state or its citizens are viewed as having the prime responsibility for the care and support of frail older people and is manifest through forms of care that are explicitly spatial in their expression. At the local level this can be viewed as a continuum that ranges from domestic to community and institutional settings, involving a shifting mix of informal and formal care-givers. Yet as the book has demonstrated, changing forms of care, particularly within advanced capitalist states, can bring new actors into the care network in new places that are beyond both the home and the institution. However, the landscape of care is also subject to wider macro-influences. So, whilst unpacking the shifting nature of care at the micro-scale is critically important, it is not sufficient. Firstly, we need to remain alert to the ways in which the macro-economic environment acts to influence the availability of formal and informal care support, impacting on differing cultural norms of rights and responsibilities within the care-giving experience. Secondly, we need to recognise that globalisation, shifting work and family structures, changing notions of community, new technologies and their ability to 'shrink distance' are all adding new layers of complexity to our attempts to understand what constitutes proximal and distance care and what this means in terms of caring *for* and caring *about*.

While particularistic forms of care-giving employ explicitly spatial expressions of care, by drawing on contemporary developments in the UK the book has drawn attention to the growing complexity of the care-giving relationship. The interconnectedness between formal and informal caring within both formal and informal sites of care has become increasingly porous. Teasing out those caring relationships that occur within public and private space and transitions between them, as well as understanding the importance of place within these relationships is crucial if we are to successfully identify: firstly, those aspects of care that act to empower and dis-empower informal carers and care-recipients; and secondly, those coping strategies that enable informal carers to continue to play an active role in the lives of care-recipients (in both formal and informal care settings) and those which facilitate older people's ability to successfully age in place.

Whilst earlier work on care within the domestic home has provided a useful framework for exploring the complex power dynamics of care, it is argued here, that in highlighting the increased porosity of the boundaries, not just between formal and informal care but also between home and institution, proximity and distance, it is possible to tease out a more nuanced understanding of the importance of people and places in the construction and delivery of care to frail older people.

Other aspects of the landscape of care however, would benefit from further analysis. It has not been possible in this book to cover all spatial aspects of informal care-giving. Some issues have been only lightly touched upon whilst others have not been addressed at all. In part, this is a consequence of the recognition of the complexity of the landscape of care that older people inhabit, but it also reflects gaps in existing geographical research around the topic. With one or two exceptions, for example Robson's (2000) and Robson et al.'s (2006) work on child carers in Sub-Saharan Africa, there is little geographical work on the important issue of child carers – particularly in western settings. In-depth understandings of how gendered (particularly male) differences in care-giving are spatially manifest are also under-represented, as are issues of ethnicity and care. Whilst the spatial aspects of sexuality and care *have* been addressed by commentators such as Brown (1996), care for lesbian and gay older people is also notable by its absence. These represent just a few aspects of the landscape of care that future research could usefully address – others, I am sure, will go on to identify new and exciting avenues of research in this field. Importantly, this book has focused on one particular aspect of the landscape of care – that impacting on older people and their informal caregivers. How the landscape of care is manifest for other groups of individuals will of course differ, this requires a 'mapping out' of the complex spatialities that are entailed in care with and for other groups. Critically, it is the way in which geographers are able to contribute new, spatialised understandings of care and care relationships that mark out their distinctive contribution to the field of care – one which this book has sought to augment.

Bibliography

Abel, E.K. (1989), 'Family care of the frail elderly: Framing an agenda for change,' *Women's Studies Quarterly* 1, 75.

Abeykoon, P. (2002), *Long-Term Care – Sri Lanka Case Study* (Geneva: WHO).

Allen, G. and Crow, G. (1989), 'Introduction' in Allen, G. and Crow, G. (eds) *Home and Family: Creating the Domestic Sphere* (London: Macmillan).

Alm, A. et al. (2004), 'A cognitive prosthesis and communication support for people with dementia,' *Neuropsychological Rehabilitation* 14:18, 117-134.

Almberg, B., Grafström, M. and Winblad, B. (1997), 'Major strain and coping strategies as reported by family members who care for aged demented relatives,' *Journal of Advanced Nursing* 26, 683-691.

Andrews, G.J. and Phillips, D.R. (2002), 'Changing local geographies of private residential care for older people 1983-1999: Lessons for social policy in England and Wales,' *Social Science and Medicine* 55, 63-78.

Aneshensel, C., Pearlin, L., Mullan, J., Zarit, S. and Whitlach, C. (1995), *Profiles in Caregiving: The Unexpected Career* (San Diego, CA: Academic Press).

Angus, J., Kontos, P.C., Dyck, I., McKeever, P. and Poland, B. (2005), 'The personal significance of home: Habitus and the experience of receiving long-term home care,' *Sociology of Health and Illness* 27, 161-187.

Apt, N.A. (2002), 'Ageing and the changing role of the family and the community: An African perspective,' *International Social Security Review* 55, 47-56.

Apt, N.A. (2004), Speech to the World Granny Conference on Ageing and Development (November 2004) (Amsterdam: Netherlands).

Arber, S. and Gilbert, N. (1989), 'Men: The forgotten carers', *Sociology* 23:1, 111-118.

Arber, S. and Ginn, J. (1990), 'The meaning of informal care: Gender and the contribution of elderly people,' *Aging and Society* 10, 429-454.

Arber, S. and Ginn, J. (1992), 'Class and caring: A forgotten dimension,' *Sociology* 26:4, 619-634.

Argylle, E. (2004), 'Age-old reasons to pay in cash,' *Community Care* 1549, 21-22.

Arksey, H. and Glendinning, C. (2007), 'Choice in the context of informal care-giving', *Health and Social Care in the Community* 15:2, 165-175.

Arksey, H. and Hirst, M. (2005), 'Unpaid carers' access to and use of primary care services,' *Primary Health Care Research and Development* 6:2, 101-16.

Askham, J., Briggs, K., Norman, I. and Redfern, S. (2007), 'Care at home for people with dementia: As in a total institution?' *Ageing and Society* 27 3-24.

Augé, M. (1995), *Non-place: Introduction to an Anthropology of Supermodernity* (translated by Howe, J.) (London: Verso).

Averill, J.R. (1996), 'The acquisition of emotions during adulthood', in Harré, R. (ed.) *The Social Construction of Emotion* (Oxford: Blackwell).

Baker, C.P.M. (1992), *Formal and Informal Support Usage of Male and Female Care-givers* (Unpublished MA Thesis) (Auckland New Zealand: University of Auckland).

Bamford, C. and Bruce, E. (2000), 'Defining the outcomes of community care: The perspectives of older people with dementia and their carers,' *Ageing and Society* 20, 543-570.

Barlow, J., Bayer, S. and Curry, R. (2005), 'Flexible homes, flexible care, inflexible organisations? The role of telecare in supporting independence,' *Housing Studies* 20:3, 441-456.

Barrientos, A. (1997), 'The changing face of pensions in Latin America: Design and prospects of individual capitalization pension plans,' *Social Policy and Administration* 31, 336-353.

Bayer, S., Barlow, J. and Curry, R. (2007), 'Assessing the impact of a care innovation: Telecare,' *Systems Dynamic Review* 23:1, 61-80.

BBC News (2003), *Census Highlights Carers' Vital Role*, 13 February. Available at: http://www.bbc.co.uk/1/hi/health/2757393.stm.

Beck, U. (2000), *What is Globalization?* (Cambridge: Policy).

Belgrave, M. and Brown, L. (1997), *Beyond a Dollar Values: Informal Care and the Northern Region Case Management Study* (New Zealand: North Health, Massey University and Waitemata Health).

Bell, D. and Bowes, A. (2006), *Lessons from the Funding of Long-term Care in Scotland*, Joseph Rowntree Foundation: London. Available at: http://www.jrf.org.uk/node/2369, accessed 12 February 2009.

Bettio, F. and Plantenga, J. (2004), 'Comparing Care Regimes in Europe', *Feminist Economics* 10:1, 85-113.

Biggs, S. (2004), 'Age, gender, narratives and masquerades', *Journal of Ageing Studies* 18, 45-58.

Blakemore, K. (2000), 'Health and social care needs in minority communities: An over-problematized issue?' *Health and Social Care in the Community* 8:1, 22-30.

Blunt, A. (2005), 'Cultural geography: Cultural geographies of home,' *Progress in Human Geography* 29:4, 505-515.

Blunt, A. and Varley, A. (2004), 'Geographies of home,' *Cultural Geographies* 11, 3-6.

Blythe, M.A., Monk, A.F. and Doughty, K. (2005), 'Socially dependable design: The challenge of ageing populations for HCI,' *Interacting with Computers* 17, 689.

Bondi, E. (1998), 'Gender, class and urban space: Public and private space in contemporary urban landscapes', *Urban Geography* 19: 160-85.

Bonomi, A.E., Anderson, M.L., Reid, R.J., Carrell, D., Fishman, P.A., Rivara, F.P. and Thompson, R.S. (2007), 'Intimate partner violence in older women', *The Gerontologist* 47:1, 34-41.

Bornat, J., Dimmock, B., Jones, D. and Peace, S. (1999), 'Stepfamilies and older people: Evaluating the implications of family change for an ageing population,' *Ageing and Society* 19, 239-261.

Bowers, B. (1988), 'Family perceptions of care in a nursing home,' *Gerontologist* 28, 361-368.

Bowes, A. and McColgan, G. (2006), *Smart Technology and Community Care for Older People: Innovation in West Lothian* (Edinburgh. Scotland: Age Concern Scotland).

Bradley, E.H., Curry, L.A., McGraw, S.A., Webster, T.R., Kasl, S.V. and Andersen, R. (2004), 'Intended use of informal long-term care: The role of race and ethnicity,' *Ethnicity and Health* 9:1, 37-54.

Brown, G. (1985), 'The discovery of expressed emotion: Induction of deduction?' In Leff, J. and Vaughn, G. (eds) *Expressed Emotion in Families: Its Significance for Mental Illness* (New York: Guilford Press) 7-25.

Brown, H. and Smith, H. (1993), 'Women caring for people: The mismatch between rhetoric and women's reality?' *Policy and Politics* 21, 185-193.

Brown, M. (2003), 'Hospice and the spatial paradox of terminal care,' *Environment and Planning A* 35, 833-851.

—— (2004), 'Between neoliberalism and cultural conservatism: Spatial divisions and multiplications of hospice labor in the United States,' *Gender, Place and Culture* 11:1, 67-82.

Brown, S.J. (2003), 'Guest editorial: Next generation telecare and its role in primary and community care,' *Health and Social Care in the Community* 11: 6, 459-462.

Burns, A. (2000), 'The burden of Alzheimer's disease,' *International Journal of Neuropsychopharmacology* 3: S31-S38.

Bywaters, P. and Harris, A. (1998), 'Supporting carers: Is practice still sexist?' *Health and Social Care in the Community* 6:6, 458-463.

Cabrera, G., Giraldez, M. and Rodriguez, C. (2005), 'The role of ambient intelligence in the social integration of the elderly in ambient intelligence,' in G. Riva, F. Vatalaro, F. Davide and M. Alcañiz (eds) *Ambient Intelligence* (Lancaster: IOS Press).

Caldock, K. (1992), 'Domiciliary services and dependency: A meaningful relationship?' In Laczko, F. and Victor, C. (eds) *Social Policy and Elderly People: The Role of Community Care* (Aldershot: Avebury) 96-111.

Call, K.T., Finch, M.A., Huck, S.M. and Kane, R.A. (1999), 'Caregiver burden from a social exchange perspective: Caring for older people after hospital discharge,' *Journal of Marriage and the Family* 61:3, 688-699.

Campbell, L.D. and Martin-Matthews, A. (2000), 'Primary and proximate: The importance of coresidence and being primary provider of care for men's filial care involvement,' *Journal of Family Issues* 21, 1006-1030.

Campbell, L.D. and Martin-Matthews, A. (2004), 'The gendered nature of men's filial care,' *Journal of Gerontology* 58B:6, 350-358.

Carmichael, F. and Charles, S. (2003), 'The opportunity cost of informal care: Does gender matter?' *Journal of Health Economics* 22, 781-803.

Chae, Y.M., Lee, J.H., Ho, S.H., Kim, H.J., Jun, K.H. and Won, J.U. (2001), 'Patient satisfaction with telemedicine in home health services for the elderly,' *International Journal of Medical Informatics* 61, 167-73.

Chamberlayne, P. and King, A. (1997), 'The biographical challenge of caring,' *Sociology of Health and Illness* 19, 5, 601-621.

Charlesworth, A., Wilkin, D. and Durie, A. (1984), *Carers and Services: A Comparison of Men and Women Caring for Dependent Elderly People* (Manchester: EOC).

Commission for Social Care Inspection (2005), *The State of Social Care in England 2004-2005.* London: CSCI. Available at: www.csci.org.uk/publications/national_reports/state_social_care_summary.htm, accessed on 14 March 2009.

Commission for Social Care Inspection (2009), *The State of Social Care in England 2007-2008*, London: CSCI. Available at: http://www.csci.org.uk/PDF/SOSC08%20Summary%2008_Web.pdf, accessed 12 February 2009.

Connidis, I. A. and Kemp, C.L. (2008), 'Negotiating actual and anticipated parental support: Multiple sibling voices in three-generation families,' *Journal of Aging Studies* 22, 229-238.

Conradson, D. (2003), 'Spaces of care in the city: The place of a community drop-in centre,' *Social and Cultural Geography* 4, 507-25.

—— (2003), 'Geographies of care: Spaces, practices, experiences,' *Social and Cultural Geography* 4:4, 451-454.

—— (2005), 'Landscape, care, and the relational self: Therapeutic encounters in rural England,' *Health and Place* 11:4, 337-348.

Cooper, C., Selwood, A. and Livingston, G. (2008), 'The prevalence of elder abuse and neglect: A systematic review,' *Age and Ageing* 37: 151-160.

Cowan, D. and Turner-Smith, A. (1999), 'The role of assistive technology in alternative models of care for older people' (Appendix 4), in Sunderland (ed.) *With Respect to Old Age: Long Term Care – Rights and Responsibilities: Research Volume 2: Alternative Models of Care for Older People*, Royal Commission on Long Term Care (London: HMSO).

Croucher, K. (2006), *Making the Case for Retirement Villages* (London: Joseph Rowntree Foundation).

Cruz-Saco, M.A. and Mesa-Lago, C. (eds) (1998), *Do Options Exist? The Reform of Pension and Health Care Systems in Latin America* (Pittsburgh, PA: University of Pittsburgh Press).

Cutchin, M.P. (2003), 'The process of mediated aging-in-place: A theoretically and empirically based model,' *Social Science and Medicine* 57, 1077-1090.

Dahlberg, L., Demack, S. and Bambra, C. (2007), 'Age and gender of informal carers: A population-based study in the UK,' *Health and Social Care in the Community* 15:5, 439-445.

Daly, M. and Lewis, J. (2000), 'The concept of social care and the analysis of contemporary welfare states,' *British Journal of Sociology* 51: 2, 281-298.

Dancy, J. Jr. and Ralston, P.A. (2002), 'Health promotions and Black elders: Subgroups of greatest need,' *Research on Ageing* 24: 218-242.

Danermark, B. and Ekstrom, M. (1990), 'Relocation and health effects on the elderly: A commented research review,' *Journal of Sociology and Social Welfare* 17:1, 25-49.

Dannefer, D., Stein, P., Siders, R. and Patterson, R.S. (2008), 'Is that all there is? The concept of care and the dialectic of critique,' *Journal of Aging Studies* 22, 101-108.

Davidson, J. and Milligan, C. (2004), 'Embodying emotion, sensing space: Introducing emotional geographies,' *Social and Cultural Geographer* 5:4, 523-532.

Davidson, J., Smith, M. and Bondi, L. (eds) (2005), *Emotional Geographies* (Aldershot: Ashgate).

Davies, C. (1995), 'Competence versus care – Care work revisited,' *Acta Sociologica* 38: 1, 17-31.

Davis, F. (1979), *Yearning for Yesterday* (New York: Free Press).

Davis, S. and Nolan, M. (2004), 'Making the move: Relatives' experiences of the transition to a care home,' *Health and Social Care in the Community* 12:6, 517-526.

Day, R. (2008), 'Local environments and older people's health: Dimensions from a comparative qualitative study in Scotland,' *Health and Place* 14, 299-312.

De Certeau, M., Giard, L. and Mayol, P. (1998), *The Practice of Everyday Life Volume 2: Living and Cooking* (Minneapolis: University of Minnesota Press).

Dear, M. and Taylor, M. (1982), *Not On Our Street: Community Attitudes to Mental Healthcare* (London: Pion).

Dear, M. and Wolch, J. (1987), *Landscapes of Despair – From Deinstitutionalisation to Homelessness* (Cambridge: Polity Press).

Demeris, G., Rantz, M.J., Aud, M.A., Marek, K.D., Tyrer, H.W., Skubic, M. and Hussam, A.A. (2004), 'Older adults' attitudes towards and perceptions of "smart home" technologies: A pilot study,' *Medical Informatics and the Internet in Medicine* 29, 87-94.

Demeris, G., Speedie, S.M., Finkelstein, S. (2001), 'Change of patients' perception of telehomecare,' *Telemed J E Health* 7:3, 241-8.

Denihan, A., Brice, I., Coakley, D. and Lawlor, B. (1998), 'Psychiatric morbidity in cohabitants of community-dwelling elderly depressives,' *International Journal of Geriatric Psychiatry* 13: 691-694.

Department of Health (1999), *Caring About Carers: A National Strategy for Carers* (London: Department of Health).

—— (2008), *Carers at the Heart of 21st-century Families and Communities* (London: HMSO).

—— (2008a) *The Preventative Technology Grant*, www.dh.gov.uk/en/ Publicationsandstatistics/Publications/PublicationsPolicyAndGuidance/ Browsable/DH_5464107, accessed November 2008.

Dickinson, A., Goodman, J., Syme, A., Eismar, R., Tiwari, L., Mival, O. and Newelli, A. (2003), 'Domesticating technology: In-home requirements gathering with frail older people,' in *A Utopia 10th International Conference on Human – Computer Interaction HCI.* 22-27 June, Crete, Greece.

Domènech, M. and Tirado, F. (1997), 'Rethinking institutions in societies of control,' *The International Journal of Transdisciplinary Studies*, 1(1), http:// www.geocities.com/Paris/Rue/8759/extituciones.html.

Domènech, M., López, D., Callen, B. and Tirado, F.J. (2006), 'Elder people and artefacts: A problem of intimacy,' Paper presented at the EASST Conference August 2006: *Reviewing Humanness: Bodies, Technologies and Spaces*, (Switzerland: University of Lausanne).

Dyck, I., Kontos, P., Angus, J. and McKeever, P. (2005), 'The home as a site for long-term care: Meanings and management of bodies and spaces,' *Health and Place* 11, 173-185.

Eborall, C. and Garmesan, K. (2001), *Desk Research on Recruitment and Retention in Social Care and Social Work* (COI Communications for the Department of Health, London).

Edvarsson, D. (2005), *Atmosphere in Care Settings: Towards a Broader Understanding of the Phenomenon* (unpublished Dissertation) (Umeå, Sweden: Department of Nursing, Umeå University).

Engstrom, M., Lunggren, B., Linqvist, R. and Carlsson, M. (2005), 'Staff perceptions of job satisfaction and life situation before and 6 and 12 months after increased information technology support in Dementia care,' *Journal of Telemedicine and Telecare* 11:6, 304-309.

Esping-Andersen, G. (1990), *The Three Worlds of Welfare Capitalism* (Cambridge: Polity Press).

—— (1999), *Social Foundations of Post-industrial Economies* (Oxford: Oxford University Press).

Esping-Anderson, G., Gallie, D., Hemerijik, A. and Myles, J. (2001), *A New Welfare Architecture for Europe? Report to the Belgian Presidency of the EU* (Brussels: CEC).

—— (2002) *Why we Need a New Welfare State* (Oxford: Oxford University Press).

Essén, A. (2008), 'The two facets of electronic care surveillance: An exploration of the views of older people who live with monitoring devices,' *Social Science and Medicine* 67, 128-136.

Evans, S. and Vallelly, S. (2008), *Social Well-being in Extra Care Housing* (London: Joseph Rowntree Foundation).

Exley, C. and Allen, D. (2007), 'A critical examination of home care: End of life care as an illustrative case', *Social Science and Medicine* 65, 2317-2327.

Fagan, C. and Burchell, B. (2002), *Gender, Jobs and Working Conditions in the European Union*, European Foundation for the improvement of Living and Working Conditions, Dublin, Ireland. Available as a PDF at: http://eric.ed.gov/ ERICWebPortal/custom/portlets/recordDetails/detailmini.jsp?_nfpb=true&_ &ERICExtSearch_SearchValue_0=ED475394&ERICExtSearch_ SearchType_0=no&accno=ED475394.

Fennell, G. (1982), *Social Interaction in Grouped Dwellings for the Elderly in Newcastle-upon-Tyne* (unpublished PhD Thesis) (Newcastle-upon-Tyne: University of Newcastle-upon-Tyne).

Ferraro, K.F. (1983), 'The health consequences of relocation among the aged in the community,' *Journal of Gerontology* 38:1, 90-96.

Finch, J. (1987), 'Whose responsibility? Women and the future of family care,' in Allen, I., Wicks, M., Finch, J. and Leat, D. (eds) *Informal Care Tomorrow* (London: PSI) 22-41.

Fleming, G. and Taylor, B.J. (2007), 'Battle on the home care front: Perceptions of home care workers of factors influencing staff retention in Northern Ireland,' *Health and Social Care in the Community* 15:1, 67-76.

Ford, R.G. and Smith, G.C. (2008), 'Geographical and structural change in nursing care provision for older people in England, 1993-2001,' *Geoforum* 39, 483-498.

Friedewald, M. and Costa, D. (2003), 'Science and technology roadmapping: Ambient intelligence in everyday life (AmI@Life)', *Joint Research Centre/ Institute for Prospective Technological Studies – European Science and Technology Observatory (JRC/IPTS-ESTO)*, Fraunhofer Institute Systems and Innovation Research ISI Institute for Prospective Technology Studies IPTS.

Froggatt, K. (2001), 'Life and death in English nursing homes: Sequestration or transition?' *Ageing and Society* 21, 319-332.

Fu Hua and Xue Di (2002), *Long Terms Care Case Study: China* (Geneva: WHO).

Fyfe, N. and Milligan, C. (2003), 'Out of the shadows: Exploring contemporary geographies of the welfare voluntary sector,' *Progress in Human Geography* 27:4, 397-413.

Ge Li and Shu Langen (2001), 'Making the transition from Family Support for the Elderly to Social Support for the Elderly,' *Chinese Sociology and Anthropology* 34:1, 35-48.

General Household Survey (2001), *General Household Survey 2001* (London: Office of National Statistics).

Gerrish, K. (2001), 'The nature and effect of communication difficulties arising from interactions between district nurses and South Asian patients and their carers,' *Journal of Advanced Nursing* 33:5, 566-574.

Getzell, M. and Feigenbaum, P. (2003), 'A case report: Implementing a nurse telecare program for treating depressions in primary care,' *Psychiatric Quarterly* 74:1, 61-73.

Giggs, J. (1973), 'The distribution of schizophrenics in Nottingham', *Transactions of the Institute of British Geographers* 59, 55-76.

Gilhooly, M. and Whittick, A. (1989), 'Expressed Emotion in care-givers of the dementing elderly,' *British Journal of Medical Psychology* 62, 265-272.

Gilliatt, S., Fenwick, J. and Alford, D. (2000), 'Public services and the consumer: Empowerment or control?' *Social Policy and Administration* 34, 333-349.

Goffman, E. (1961), *Asylums: Essays on the Social Situation of Mental Patients and other Inmates* (Harmondsworth: Penguin Books).

Golant, S. (1999), 'The promise of assisted living as a shelter and care alternative for frail American elders,' in Schwarz, B. and Brent, R. (eds) *Aging, Autonomy and Architecture: Advances in Assisted Living* (Baltimore: The John Hopkins University Press).

Goldani, A.M. (1990), 'Changing Brazilian families and the consequent need for public policy,' *International Social Science Journal* 42, 523-537.

Gollins, T. (2004), *Male Carers: A Study of the Inter-relationships between Caring, Male Identity and Age*, www.shef.ac.uk/socst/Shop/gollins.pdf.

Gott, M., Seymour, J., Bellamy, G., Clark, D. and Ahmedzhai, S. (2004), 'Older people's views about home as a place of care at the end of life,' *Palliative Medicine* 18, 460-467.

Graham, H. (1983), 'Caring: A labour of love', in Finch, J. and Groves, D. (eds) *A Labour of Love: Women, Work and Caring* (London: Routledge and Kegan Paul).

—— (1991), 'The concept of caring in feminist research: The case of domestic service,' *Sociology* 25: 61-78.

Graham, P.M., Nancarrow, S.B., Parker, H., Phelps, K. and Regend, E. (2005), 'Place, policy and practitioners: On rehabilitation, independence and the therapeutic landscape in the changing geography of care provision to older people in the UK,' *Social Science and Medicine* 61, 1893-1904.

Griffiths, J. (1998), 'Meeting personal hygiene needs in the community: A district nursing perspective on the health and social care divide,' *Health and Social Care in the Community* 6:4, 234-240.

Groger, L. (1996), 'A Nursing Home can be a home,' *Journal of Aging Studies* 9, 137-153.

Grundy, E. and Henretta, J.C. (2006), 'Between elderly parents and adult children: A new look at the intergenerational care provided by the "sandwich generation,"' *Ageing and Society* 26:5, 707-722.

Gustke, S.S., Balch, D.C., West, V.L. and Rogers, L.O. (2000), 'Patient satisfaction with Telemedicine,' *Telemed Journal* 6:1, 5-13.

Hall, E. and Kearns, R. (2001), 'Making space for the "intellectual" in geographies of disability,' *Health and Place* 7:3, 237-246.

Hammer, E. and Österle, A. (2003), 'Welfare state policy and informal long-term care-giving in Austria: Old gender divisions and new stratification processes among women', *Journal of Social Policy* 32:1, 37-53.

Hargreaves, P., Kirsch, A.J. and Heywood, P. (2004), *Money Well Spent: The Effectiveness and Value of Housing Adaptations* (Bristol: Policy Press).

Hargreaves, P., Kirsch, A.J., Robinson, P., Green, A., Mann, E.Z., Bayer, S., Barlow, J. and Curry, R. (2007), 'Assessing the impact of a care innovation: Telecare,' *System Dynamics Review* Vol. 23: 1, 61-80.

Harvey, D. (1996), *Justice, Nature and the Geography of Difference* (Cambridge, Mass: Blackwell).

Help the Aged (2009), 'Sheltered housing "in crisis" due to loss of site wardens,' *Community Care* 03075508, 29 January 2009, Issue 1755. Available at: http://www.communitycare.co.uk/Articles/2009/01/23/110520/, accessed 10 March 2009.

Henwood, M. (1990), *Community Care and Elderly People* (London: Family Policy Studies Centre).

Hepworth, M. (1998), 'Ageing and emotions,' in Bendelow, G. and Williams, S.J. (eds) *Emotions in Social Life: Critical Themes and Contemporary Issues* (London: Routledge) 173-189.

Herzberg, A., Ekman, S.-L., Axelsson, K. (2001), 'Staff activities and behaviour are the source of many feelings: Relatives' interactions and relationships with staff in nursing homes', *Journal of Clinical Nursing* 10: 380-388.

Hirst, M. (2001), 'Trends in informal care in Great Britain during the 1990s,' *Health and Social Care in the Community* 9:6, 348-357.

Hirst, M. (2003), 'Caring-related inequalities in psychological distress in Britain during the 1990s,' *Journal of Public Health Medicine* 25:4, 336-343.

Hirst, M. (2005), 'Carer distress: A prospective, population-based study,' *Social Science and Medicine* 61, 697-708.

HM Government (2008), *The Case for Change – Why England Needs a New Care and Support System*, DH Publications online, London. Available at: http://www.dh.gov.uk/en/Publicationsandstatistics/Publications/index.htm, accessed 10 February 2009.

Hochschild, A. (1979), 'Emotion work, feeling rules, and social structure,' *The American Journal of Sociology* 85:3, 551-575.

Hochschild, A. (1993), Preface in Fineman, S. (ed.) *Emotion in Organisation* (London: Sage).

Hockey, J. (1990), *Experiences of Death: An Anthropological Account* (Edinburgh: Edinburgh University Press).

Hogenbirk, J., Liboiron-Grenier, L., Pong, R. and Young, N. (2005), *How Can Telehomecare Support Informal Care? Examining What is Known and Exploring the Potential*, Final Report to Home and Continuing Care Policy Unit, Health Canada, www.hc-sc.gc.ca/hcs-sss/alt_formats/hpb-dgps/pdf/pubs/2005-tele-home-domicile/2005-tele-home-domicile-eng.pdf, accessed 12 March 2009.

Holroyd, E. (2003), 'Hong Kong Chinese family caregiving: Cultural categories of bodily order and the location of self,' *Health Research* 13:2, 158-170.

INSTAT (2004), *Migration in Albania* (Tirana: Instituti i Statistikës).

Iwarsson, S., Wahl, H.W., Nygren, C., Oswald, F., Sixsmith, A., Sixsmith, J., Sze'man, Z. and Tomsone, S. (2007), 'Importance of the home environment for healthy aging: Conceptual and methodological background of the European ENABLE–AGE Project,' *The Gerontologist* 47:1, 78-84.

Jönsson, I. (2003), 'Policy perspectives on Changing Intergenerational relations,' *Social Policy and Society* 2, 241-248.

Joseph, A.E., Kearns, R.A. and Moon, G. (2009), 'Recycling former psychiatric hospitals in New Zealand: Echoes of deinstitutionalisation and restructuring,' *Health and Place* 15:1, 79-87.

Katz, C. and Monk, J. (eds) (1993), *Full Circles: Geographies of Women over the Life Course* (London: Routledge).

Keeley, B. and Clarke, M. (2002), *Carers Speak Out Project: Report on Finding and Recommendations* (London: The Princess Royal Trust for Carers).

Keister, K.J. (2006), 'Predictors of self-assessed health, anxiety, and depressive symptoms in nursing home residents at week 1 postrelocation,' *Journal of Aging and Health* 18:5, 722-742.

Kellet, U. (1999), 'Searching for new possibilities to care: A qualitative analysis of family caring involvement in nursing homes,' *Nursing Inquiry* 6: 9-16.

Kim Ik Ki and Maeda Daisauku (2001), 'A comparative study on socio-demographic changes in long-term health care needs of the elderly in Japan and South Korea,' *Journal of Cross-Cultural Gerontology* 16, 237-255.

Koffman, J.S. and Higginson, I.J. (2003), 'Fit to care? A comparison of informal caregivers of first-generation Black Caribbeans and White dependants with advanced progressive disease in the UK,' *Health and Social Care in the Community* 11:6, 528-536.

Kontos, P.C. (1998), 'Resisting institutionalization: Constructing old age and negotiating home,' *Journal of Aging Studies* 12:2, 167-185.

Laflamme, M.R., Wilcox, D.C., Sullivan J., Shadow, G. et al. (2005), 'A Pilot Study of Usefulness of Clinician-Patient Videoconferencing for Making Routine Medical Decisions in the Nursing Home,' *Journal of the American Geriatrics Society* 53:8, 1380.

Laing and Buisson (1992), *Laing's Review of Private Healthcare 1992* (London: Laing and Buisson).

Lan, P.C. (2001), *Subcontracting Filial Piety: Elder Care in Dual-earner Chinese Immigrant Households in the Bay Area*, Working Paper No. 21 (Berkeley, CA, USA: Centre for Working Families, University of California).

Lansley, P., McCreadie, C. and Tinker, A. (2004), 'Can adapting the homes of older people and providing assistive technology pay its way?' *Age and Ageing* 33, 576.

Lau, B.W.K. and Pritchard, C. (2001), 'The suicide of older people in Asian societies: An international comparison,' *Australasian Journal of Ageing* 20:4, 196-202.

Lawson, V. (2007), 'Geographies of Care and Responsibility,' *Annals of the Association of American Geographers* 91:1, 1-11.

Lawton, J. (1998), 'Contemporary hospice care: The sequestration of the unbounded body and dirty dying,' *Sociology of Health and Illness* 20:2, 121-143.

Lawton, J. (2000), *The Dying Process: Patients' Experiences of Palliative Care* (London: Routledge).

Lee-Treweek, G. (1996), 'Emotion work, order and emotional power in care assistant work,' in James, V. and Gabe, J. (eds) *Health and the Sociology of Emotions* (Oxford: Blackwell).

Levitt, P. (2001), *The Transnational Villagers* (Berkeley, California: University of California Press).

Lewis, J. (2001), 'Older people and the health-social care boundary in the UK: Half a century of hidden policy conflict,' *Social Policy and Administration* 35:4, 343-359.

Lewis, J. (2002), 'Gender and welfare state change,' *European Societies* 4, 331-356.

Lianos, M. (2003), 'Social control after Foucault,' *Surveillance and Society* 1:3, 412-430.

Litt, J.S. and Zimmerman, M.K. (2003), 'Global perspectives on gender and carework: An introduction,' *Gender and Society* 17, 156-165.

Livingston, J. (2002), 'Reconfiguring old age: Elderly women and concerns over care in Southeastern Botswana,' *Medical Anthropology* 22, 205-231.

Lloyd-Sherlock, P. (2000), 'Old age and poverty in developing countries: New policy challenges,' *World Development* 28:12, 2157-2168.

Lloyd-Sherlock, P. (2003), 'Financing health services for pensioners in Argentina: A salutary tale,' *International Journal of Social Welfare* 12, 24-30.

Lloyd-Sherlock, P. and Locke, C. (2008), 'Vulnerable relations: Lifecourse, wellbeing and social exclusion in Buenos Aires, Argentina,' *Ageing and Society* 28, 1177-1201.

Longman, P., Ackerman, E., Boroughs, D., Fang, B., Hart, D. and Myers, B. (1999), 'The world turns grey,' *US News and World Report* 126:8, 1-8.

López, D. (2006), 'La Teleasistencia Domiciliaria como extitución. Análisis de las nuevas formas espaciales del cuidado', in Tirado, F.J. and Domènech, M. (eds) *Lo social y lo virtual. Nuevas formas de control y transformación social* (Barcelona: Editorial UOC) 60-78.

Lopez, D. and Domenech, M. (2006), 'Risky houses for independent elderly: The question concerning immediacy in a home telecare service,' paper presented at MEDUSE workshop, Lancaster University, Lancaster, UK.

—— (2008), 'On inscriptions and ex-inscriptions: The production of immediacy in a home telecare service,' *Environment and Planning D*, 26, 663-675.

Lundell, J. and Morris, M. (2004), 'Tales, tours, tools, and troupes: A tiered research method to inform ubiquitous designs for the elderly,' 18th Annual Conference of the British-HCI-Group, Leeds Metropolitan University, UK.

Lyon, D. (2006), *Theorizing Surveillance: The Panopticon and Beyond* (Devon, UK: Willan Publishing).

Lyon, D. (2007), *Surveillance Studies: An Overview* (Cambridge, MA: Polity Press).

Machado, L. (2001), *Elder Abuse in Brazil* (Geneva: WHO).

Magnusson, L. and Hanson, E.J. (2003), 'Ethical issues arising from a research, technology and development project to support frail older people and their family carers at home,' *Health and Social Care in the Community* 431-439.

Magnusson, L., Hanson, E. and Borg, M. (2004), 'A literature review study of Information and Communication Technology as a support for frail older people living at home and their family carers,' *Technology and Disability* 16, 223.

Maher, J. and Green, H. (2002), *Carers 2000* (London: The Stationery Office).

Mahoney, D.M.F., Tarlow, B., Jones, R.N., Tennsted, S. and Kasten, L. (2001), 'Factors affecting the use of a telephone-based intervention for caregivers of people with Alzheimer's disease,' *Journal of Telemedicine and Telecare* 7:3, 139-148.

Manthorpe, J. and Iliffe, S. (2005), 'Respite care and short-term support: New forms of an old idea?' *Nursing Older People* 17, 15-16.

Manthorpe, J., Iliffe, S. and Eden, A. (2001), 'Testing Twigg and Atkin's typology of caring: A study of primary care professionals' perceptions of dementia care using a modified focus group method,' *Health and Social Care in the Community* 11:6, 477-485.

Martin, E.M. and Coyle, M.K. (2006), 'Nursing protocol for telephonic supervision of clients,' *Rehabilitation Nursing* 31:2, 54-57.

Massey, D. (2002), 'Globalisation: What does it mean for geography?' *Geography* 87:4, 293-296.

Matthews, S.H. (2002), *Sisters and Brothers/Daughters And Sons: Meeting the Needs of Older Parents* (Bloomington, IN: Unlimited Publishing).

McCann, S., Ryan, A. and McKenna, H. (2005), 'The challenges associated with providing community care for people with complex needs in rural areas: A qualitative investigation,' *Health and Social Care in the Community* 13:5, 462-469.

McCreadie, C. and Tinker, A. (2005), 'The acceptability of assistive technology to older people,' *Ageing and Society* 25, 91-110 Part 1.

McDaid, D. and Sassi, F. (2001), 'The burden of informal care for Alzheimer's disease: Carers' perceptions from an empirical study in England, Italy and Sweden,' *Mental Health Research Review* 8, 34-26.

McDowell, L. (1992), 'Social divisions, income inequality and gender relations in the 1980s' in Cloke, P. (ed.) *Policy and Change in Thatcher's Britain* (Oxford: Pergamon Press) 355-378.

McGarry, J. and Simpson, C. (2008), 'Identifying, reporting and preventing elder abuse in the practice setting,' *Nursing Standard* 22:46, 49-55.

McKeever, P. (2001), '"Hitting Home": The home as a locus of long-term care,' Paper presented to the Association of American Geographers Annual Meetings, New York.

McKibbin, R. (1998), *Classes and Cultures: England 1981-1951* (Oxford: Oxford University Press).

McPherson, M. (2003), *The Nature and Role of the Extended Family in New Zealand* (Palmerston North, New Zealand: Social Policy research Centre, Massey University).

Meleis, A.I., Sawyer, L.M., Im, E.O., Hilfinger Messias, D.K. and Schumacher, K. (2000), 'Experiencing transitions: An emerging middle-range theory,' *Advanced Nursing Science* 23:1, 12-28.

Meresman, J.F., Hunkeler, E.H., William, A. and Manthorpe, J. (2004), 'Risk taking,' in Innes, A., Archibald, C. and Murphy, C. (eds) *Dementia and Social Inclusion: Marginalised Groups and Marginalised Areas of Dementia Research, Care and Practice* (London: Jessica Kingsley Press) 137-149.

Merrell, J., Kinsella, F., Murphy, F., Philpin, S. and Ali, A. (2006), 'Accessibility and equity of health and social care services: Exploring the views and experiences of Bangladeshi carers in South Wales, UK,' *Health and Social Care in the Community* 14:3, 197-205.

Meth, P. (2003), 'Entries and omissions: Using solicited diaries in geographical research,' *Area* 35:2, 195-205.

Miller, E.A. (2001), 'Telemedicine and doctor–patient communication: An analytical survey of the literature,' *Journal of Telemedicine and Telecare* 7:1, 1-17.

Milligan, C. (1999), 'Without these walls: A geography of mental ill health in a rural environment,' in Butler, R. and Parr, H. (eds) *Mind and Body Spaces: Geographies of Illness, Impairment and Disability* (London: Routledge).

—— (2000), 'Bearing the burden: Towards a restructured geography of caring,' *Area* 32, 49-58.

—— (2001), *Geographies of Care: Space, Place and the Voluntary Sector* (Aldershot: Ashgate).

—— (2003), 'Location or Dis-location: From community to long term care – The caring experience,' *Journal of Social and Cultural Geography* 4, 455-470.

—— (2004), 'Caring for older people in New Zealand: Informal carers' experiences of the transition of care from the home to residential care,' Research Report, http://www.lancs.ac.uk/fss/ihr/publications/christinemilligan/.

—— (2005), 'From home to "home": Situating emotions within the care-giving experience', *Environment and Planning A* 37:12, 2105-2120.

—— (2006), 'Caring for older people in the 21st Century: Notes from a small island,' *Health and Place* 12, 320-331.

Milligan, C., Bingley, A. and Gatrell, A. (2005), '"Healing and Feeling": The place of emotions in later life,' in: Davidson, J., Bondi, L. and Smith, M. (eds) *Emotional Geographies* (London: Blackwell).

Milligan, C. and Conradson, D. (eds) (2006), *Landscapes of Voluntarism: New Spaces of Health, Welfare and Governance* (Bristol: Policy Press).

Milligan, C., Kyle, R., Bondi, L., Fyfe, N.R., Kearns, R. and Larner, W. (2008), *From Placards to Partnership: The Changing Nature of Community Activism and Infrastructure in Manchester, UK and Auckland, Aotearoa New Zealand* (Institute for Health Research, Lancaster University, UK).

Milligan, C., Mort, M. and Roberts, C. (in press), 'Locating new care technologies in the home: From place to non-place? *Space and Culture,*' invited chapter editors: Domenech, M. and Schillmeier, M. *Care and the Art of Dwelling: Bodies, Technologies and Home* (Aldershot: Ashgate).

Ministry of Health (2008), *Statement of Intent 2008-2011* (New Zealand Ministry of Health, Wellington).

Moon, G. (2000), 'Risk and protection: The discourse of confinement in contemporary mental health policy,' *Health and Place* 3:1, 239-250.

Morris, J. (2004), 'Independent living and community care: A disempowering framework,' *Disability and Society* 19:5, 427-442.

Morris, M.E. (2005), 'Social networks as health feedback displays,' *IEEE Internet Computing.*

Morris, M., Lundell, J. and Dishman, E. (2004), 'Catalyzing social interaction with ubiquitous computing: A needs assessment of elders coping with cognitive decline,' *CHI* 24-29.

Morris, M., Lundell, J., Dishman, E. and Needham, B. (2003), 'New perspectives on ubiquitous computing from ethnographic study of elders with cognitive decline,' *UBICOMP* LNCS 2864, 227-242.

Mort, M., Milligan, C., Roberts, C. and Moser, I. (eds) (2008), *Ageing, Technology and Home Care: New Actors, New Responsibilities* (Paris: Presses de l'Ecole des mines).

Moss, P. (1997), 'Negotiating spaces in home environments: Older women living with arthritis,' *Social Science and Medicine* 45:1, 23-33.

Murray, U. and Brown, D. (1998), *'They Look After Their Own, Don't They?' Inspection of Community Care Services for Black and Ethnic Minority Older People*, Department of Health Social Care Group and Social Services Inspectorate. www.dh.gov.uk/assetRoot/04/08/42/86/04084286.pdf, accessed 17 March 2009.

Najafizadeh, M. (2003), 'Women's empowering carework in Post-Soviet Azerbaijan,' *Gender and Society* 17, 293-304.

National Alliance of Caregiving (2006), *The MetLife Caregiving Cost Study: Productivity Losses to U.S. Business* (Westport, CT: Metlife Mature Market Institute).

Nazroo, J. (1997), *Health of Britain's Ethnic Minorities* (London: Policy Studies Institute).

Netten, A., Darton, R., Bebbington, A. and Brown, P. (2001), 'Residential or nursing home care? The appropriateness of placement decisions,' *Ageing and Society* 21, 3-23.

NHS Health and Social Care Information Centre (2006), *Community Care Statistics 2004-2005*, available as pdf at: http://www.ic.nhs.uk/webfiles/publications/ commcare05adultengrepcssr/CommunityCareStatistics280206_PDF.pdf, accessed 25 March 2009.

Nippert-Eng, C. (1996), *Home and Work* (Chicago: University of Chicago Press).

Nissel, M. (1980), 'A greater place for family responsibility?' in Nissel, M., Maynard, A., Young, K. and Ibsen, M. (eds) *The Welfare State – Diversity and Decentralisation*, Discussion Paper 2 (London: Policy Studies Institute) 4-21.

Nolan, M. and Dellasega, C. (1999), '"It's not the same as him being at home": Creating caring partnerships following nursing home placement,' *Journal of Clinical Nursing* 8, 723-730.

Nordentoft, H.M. (2008), 'Changes in Emotion work at interdisciplinary conferences following clinical supervision in a palliative outpatient ward,' *Qualitative Health Research* 18:7, 913-927.

Nuttall, S., Blackwood, R., Bussell, B. et al. (1994), 'Financing long-term care in Great Britain,' *Journal of the Institute of Actuaries* 121, 1-68.

O'Reilly, D., Connolly, S., Rosato, M. and Patterson, C. (2008), 'Is caring associated with an increased risk of mortality? A longitudinal study,' *Social Science and Medicine* 67:8, 1282-1290.

Offen, K. (1992), 'Defining feminism: A comparative historical approach', in Bock, G. and James, S. (eds) *Beyond Equality and Difference Citizenship, Feminist Politics and Female Subjectivit* (London: Routledge).

Offer, J. (1999), 'Idealist thought, social policy and the rediscovery of informal care,' *British Journal of Sociology* 50:3, 467-488.

Office of National Statistics (2000), *Carers 2000*, www.statistics.gov.uk/releases, accessed 15 March 2009.

—— (2008), *Ageing: More Pensioners than Under-16's for First Time Ever*, http://www.statistics.gov.uk/cci/nugget.asp?ID=949, accessed 12 November 2008.

Oldman, C. (2002), 'Later life and the social model of disability: A comfortable partnership?' *Ageing and Society* 22, 791-806.

Oliver, M. (1992), 'Changing the social relations of research production,' *Disability, Handicap and Society* 7, 101-11.

Oudshoorn, N. (2006), 'Exploring and rethinking invisibility in the context of telemedicine,' Paper presented at the workshop 'Material Narratives of Technology in Society', University of Twente, Enschede, 19-21 October.

Parr, H. (2003), 'Medical geography: Care and caring,' *Progress in Human Geography* 27:2, 212-221.

Parr, H. and Philo, C. (2003), 'Rural mental health and social geographies of caring,' *Social and Cultural Geography* 4:4, 471-488.

Peace, S. (1986), 'The forgotten female: Social policy and older women,' in Phillipson, C. and Walker, A. (eds) *Ageing and Social Policy: A Critical Assessment* (Aldershot: Gower).

Peace, S. and Holland, C. (2001), 'Homely residential care: A contradiction in terms?' *Journal of Social Policy* 30:3, 393-410.

Peace, S., Holland, C. and Kellaher, L. (2006), *Environment and Identity in Later Life* (Milton Keynes: Open University Press).

Peace, S., Kellaher, L. and Willcocks, D. (1997), *Re-evaluating Residential Care* (Buckingham: Open University Press).

Percival, J. (2001), 'Self-esteem and social motivation in age-segregated settings,' *Housing Studies* 16: 6, 827-840.

Percival, J. (2002), 'Domestic spaces: Uses and meanings in the daily lives of older people,' *Ageing and Society* 22, 729-49.

Phillipson, C. (2007), 'The "elected" and the "excluded": Sociological perspectives on the experience of place and community in old age,' *Ageing and Society* 27, 321-342.

Pols, J. and Moser, I. (2009), 'Cold technologies versus warm care? On affective and social relations with and through care technologies,' *ALTER: European Journal of Disability Studies*, 159-178.

Powell, S. and Biggs, J.L. (2001), 'A Foucauldian analysis of old age and the power of social welfare,' *Journal of Aging and Social Policy* 12:2, 93-113.

Proctor, J. and Smith, D. (eds) (1999), *Geography and Ethics: Journeys in a Moral Terrain* (London: Routledge).

Pruchno, R.A. and Rose, M.S. (2000), 'The effect of long-term care environments on health outcomes,' *The Gerontologist* 40, 422-428.

Qureshi, H. and Walker, A. (1989), *The Caring Relationship: Elderly People and Their Families* (London: Macmillan).

Ramos, L. (2000), 'Aging in Brazil,' *Ageing International* Spring 2000, 58-64.

Rao Kequin, Yi Li and Liu Yuan Li (2000), 'Transactions of residents' health, change of health service needs and their influence on economic and social development,' *Chinese Health Economics* 19:10, 8-11.

Reed-Danahay, D. (2001), 'This is your home now! Conceptualising location and dislocation a dementia unit,' *Qualitative Research* 1: 47-63.

Relph, E. (1976), *Place and Placelessness* (London: Pion).

Roberts, C. and Mort, M. (2009), 'Reshaping what counts as care: Older people, work and new technologies,' *ALTER: European Journal of Disability Studies*, 138-158.

Robinson, J. and Banks, P. (2005), *The Business of Caring: King's Fund Inquiry into Care Services for Older People in London* (London: King's Fund).

Robson, E. (2000), 'Invisible carers: Young people in Zimbabwe's home-based healthcare,' *Area* 32: 56-69.

Robson, E., Ansell, N., Huber, U.S. and Gould, W.T.S. (2006), 'Young caregivers in the context of the HIV/AIDS pandemic in Sub-Saharan Africa,' *Population Space and Place* 12:2, 93-111.

Rogero-García, J., Prieto-Flores, M. and Rosenberg, M. (2008), 'Health services use by older people with disabilities in Spain: Do formal and informal care matter?' *Ageing and Society* 28, 959-978.

Romoren, T.I. (2003), 'The carer careers of son and daughter primary carers of their very old parents in Norway,' *Ageing and Society* 23:4, 471-485.

Rosato, M. and O'Reilly, D. (2006), 'Should uptake of state benefits be used as indicators of need and disadvantage?' *Health and Social Care in the Community* 14:4, 294-301.

Rowles, G.D. (1978), *Prisoners of Space? Exploring the Geographical Experiences of Older People* (Boulder, Colorado: Westview Press).

—— (1987), 'A place called home,' in Carstensen, L. and Edelstein, B. (eds) *Handbook of Clinical Gerontology* (New York: Pergamon Press).

—— (1993), 'Evolving images of place in aging and aging in place,' *Generations* 17:2, 65-51.

Rowles, G.D. and High, D.M. (1996), 'Individualizing care: Family roles in nursing home decision-making,' *Journal of Gerontological Nursing* 22: 20-25.

Rubery, J., Smith, M. and Fagan, C. (1999), *Women's Employment in Europe: Trends and Prospects* (London and New York: Routledge).

Rubinstein, R. (1989), 'The home environments of older people: A description of psychosocial processes linking person to place,' *Journal of Gerontology* 44: S45-53.

Ryan, A. and Scullion, H. (2000a), 'Family and staff perceptions of the role of families in nursing homes,' *Journal of Advanced Nursing* 32: 626-634.

Ryan, A. and Scullion, H. (2000b), 'Nursing home placement: An exploration of the experience of family carers,' *Journal of Advanced Nursing* 32: 1187-1195.

Sacco and Reza Nakhaie (2001), 'Crime: An examination of elderly and non-elderly adaptations,' *International Journal of Law and Psychiatry* 24: 305-323.

Savage, M., Bagnall, G. and Longhurst, B. (2005), *Globalization and Belonging* (London: Sage).

Serres, M. (1994), *Atlas* (Madrid: Cátedra).

Shaibu, S. and Wallhagen, M. (2002), 'Family care-giving of the elderly in Botswana: Boundaries of culturally acceptable options and resources,' *Journal of Cross-cultural Gerontology* 17, 139-154.

Shukla, R. and Brooks, D. (1996), *A Guide to Care of the Elderly* (London: HMSO).

Silk, J. (2004), 'Caring at a distance: Gift theory, aid chains and social movements,' *Social and Cultural Geography* 5:2, 229-251.

Skilbeck, J. and Payne, S. (2003), 'Emotional support and the role of clinical nurse specialists in palliative care,' *Journal of Advanced Nursing* 43:5, 521-530.

Skinner, M. and Rosenberg, M. (2006), 'Informal and voluntary care in Canada: Caught in the Act?' In Milligan, C. and Conradson, D. (eds) *Landscapes of Voluntarism: Health, Welfare and Governance* (Bristol: Policy Press) 91-114.

Skinner, M., Yantzi, N. and Rosenberg, M. (2009), 'Neither rain nor hail nor sleet nor snow: Provider perspectives on the challenges of weather for home and community care,' *Social Science and Medicine* 68, 682-688.

Smith, D.M. (1998), 'How far should we care? On the spatial scope of beneficence,' *Progress in Human Geography* 22: 15-38.

—— (2000), *Moral Geographies: Ethics in a World of Difference* (Edinburgh: Edinburgh University Press).

Smith, J. (2005), 'The uneven geography of global civil society: National and global influences on transnational association,' *Social Forces* 84:2 621-653.

Soopramanien, A., Pain, H., Stainthorpe, A., Menarini, M. and Ventura, M. (2005), 'Using telemedicine to support spinal-injured patients after discharge,' *Journal of Telemedicine and Telecare* 11, 68-S70.

Spitzer, D., Neufeld, A., Harrison, M., Hughes, K. and Stewart, M. (2003), 'Caregiving in transnational context: "My wings have been cut; where can I fly?"' *Gender and Society* 17:2, 267-286.

Sweeting, H. and Gilhooly, M. (1997), 'Dementia and the phenomenon of social death,' *Sociology of Health and Illness* 19:1, 93-117.

Swift, P. (2007), 'Champions can leap hurdles,' *Community Care* 1689, 32-33.

Szengonzi, R. (2007), 'The plight of older persons as caregivers to people infected/affected by HIV/AIDS: Evidence from Uganda,' *Journal of Cross-cultural Gerontology* 22, 339-353.

Tanner, D. (2001), 'Sustaining the self in later life: Supporting older people in the community,' *Ageing and Society* 21, 255-278.

Taylor, D.K., Bachuwa, G., Evans, J. and Jackson-Johnson, V. (2006), 'Assessing barriers to the identification of elder abuse and neglect: A community-wide survey of primary care physicians,' *Journal of the National Medical Association* 98:3, 403-404.

Team, V., Markovic, M. and Manderson, L. (2007), 'Family caregivers: Russian-speaking Australian women's access to welfare support,' *Health and Social Care in the Community* 15:5, 397-406.

ter Meulen, R. and van der Made, J. (2000), 'The extent and limits of solidarity in Dutch health care', *International Journal of Social Welfare* 9, 250-260.

Thomas, C. (2007), *Sociologies of Disability and Illness: Contested Ideas in Disability Studies and Medical Sociology* (Hampshire: Palgrave Macmillan).

Thomas, C., Morris, S. and Harman, J. (2002), 'Companions through cancer: The care given by informal carers in cancer contexts,' *Social Science and Medicine* 54, 529-544.

Tinker, A. (1999), *Ageing in Place: What Can We Learn From Each Other?* The Sixth F. Oswald Barnett Oration, www.sisr.net/events/docs/obo6.pdf, accessed 18 February 2009.

Tinker, A., McCreadie, C. and Lansley, P. (2003), 'Assistive technology: Some lessons from the Netherlands,' *Gerontechnology* 2, 332-337.

Tinker, A., McCreadie, C., Stuchbury, R., Turner-Smith, A., Cowan, D., Bialokoz, A., Lansley, P., Bright, K., Flanagan, S., Goodacre, K. and Goodacre, P.H. (2004), *At Home with AT: Introducing Assistive Technology into the Existing Homes of Older People – Feasibility, Acceptability, Costs and Outcomes* (London: Institute of Gerontology King's College London and the University of Reading).

Tivers, J. (1987), 'Women with young children: Constraints on activity in the urban environment,' in Little, J., Peake, L. and Richardson, P. (eds) *Women in Cities: Gender and the Urban Environment* (London: Macmillan) 84-97.

Townsend, P. (1957), *The Family Life of Old People* (London: Routledge).

—— (1962), *The Last Refuge: A Survey of Residential Institutions and Homes for the Aged in England and Wales* (London: Routledge and Kegan Paul).

—— (1965), 'The Aged in the welfare state: The interim report of a survey of persons aged 65 and over in Britain, 1962 and 1963,' *Occasional Papers on Social Administration* No 14, London.

Tracy, C.S., Drummond, N., Ferris, L.E., Globerman, J., Hebert, P.C., Pringle, D.M. and Cohen, C.A. (2004), 'To tell or not to tell? Professional and lay perspectives on the disclosure of personal health information in community-based dementia care,' *Canadian Journal on Ageing-Revue Canadienne Du Vieillissemen* 23, 203-215.

Tronto, J. (1993), *Moral Boundaries: A Political Argument for an Ethic of Care* (London: Routledge).

Tuan, Y.-F. (2004), 'Home,' in Harrison, S., Pile, S. and Thrift, N. (eds) *Patterned Ground: The Entanglements of Nature and Culture* (London: Reaktion Books) 164-65.

Tucker, S., Hughes, J., Burns, A. and Challis, D. (2008), 'The balance of care: Reconfiguring services for older people with mental health problems,' *Aging and Mental Health* 12:1, 81-91.

Turner, J. (2005), *Social Security Pensionable Age in OECD Countries: 1949-2035*, AARP Public Policy Institute, Washington DC. Available at: http://assets. aarp.org/rgcenter/econ/2005_16_oecd.pdf, accessed 27 October 2008.

Twigg, J. (1989), 'Models of carers: How do social care agencies conceptualise their relationship with informal carers?' *Journal of Social Policy* 18: 53-66.

—— (1997), 'Deconstructing the social bath: Help with bathing at home for older and disabled people,' *Journal of Social Policy* 26, 211-232.

—— (1999), 'The spatial ordering of care: Public and private in bathing support at home,' *Sociology of Health and Illness* 21: 381-400.

—— (2000), *Bathing – The Body and Community Care* (London: Routledge).

Ungerson, C. (ed.) (1990), *Gender and Caring: Work and Welfare in Britain and Scandinavia* (London: Harvester Wheatsheaf).

Valins, O. (2006), 'The difference of voluntarism: The place of voluntary sector care homes for older Jewish people in the United Kingdom,' in Milligan, C. and Conradson, D. (eds) *Landscapes of Voluntarism: Health, Welfare and Governance* (Bristol: Policy Press) 135-152.

Van Dullemen, C. (2006), 'Older people in Africa: New engines to society?' *NWSA Journal* 18:1, 99-105.

Verzhikovskaya, N.V., Ekhneva, T.L. and Beliy, A. (1999), 'Medico-social unit at OPD as a separate form of domiciliary medical care for the elderly disabled,' *Problems of Ageing and Longevity* 8, 335-342.

Vishnevskij, A. (1995), 'The modernisation of Russia', in Zaslavskaja, T. (ed.) *Where Goes Russia? Alternatives for Social Development* (Moscow: Aspekt).

Vitaliano, P., Young, H., Russo, J., Romano, J. and Magnato-Amato, A. (1993), 'Does expressed emotion in spouses predict subsequent problems among care-recipients with Alzheimer's disease?' *Journal of Gerontology* 48, 202-209.

Vitores, A. (2002), 'From hospital to community: Case management and the virtualization of institutions,' *Athenea Digital* 1, 1-6, http://ddd.uab.es/pub/athdig/15788946n1a13.pdf, accessed 12 November 2008.

Vullnetari, J. and King, R. (2008), '"Does your granny eat grass?" On mass migration, care drain and the fate of older people in rural Albania,' *Global Networks* 8: 2, 139-17.

Wanless, D. (2006), *Securing Good Care for Older People: Taking a Long-term View* (London: Kings Fund).

Warnes, A.M., Friedrich, K., Kellaher, L. and Torres, S. (2004), 'The diversity and welfare of older migrants in Europe,' *Ageing and Society* 24, 307-326.

Weardon, A., Tarrier, N., Barrowclough, C., Zastowny, T. and Rahill, A. (2000), 'A review of expressed emotion research in health care,' *Clinical Psychology Review* 20:5, 633-666.

Wenger, C. (2001), 'Myths and realities of ageing in rural Britain,' *Ageing and Society* 21, 117-130.

Wiles, J. (2003), 'Daily geographies of care-givers,' *Social Science and Medicine* 57, 1307-1325.

——— (2005), 'Home as a new site of care provision and consumption,' in Andrews, G. and Phillips, D. (eds) *Ageing and Place: Perspectives, Policy and Practice* (London: Routledge) 79-97.

Wiles, J., Allen, R., Palmer, A., Hayman, K., Keeling, S. and Kerse, N. (2009), 'Older people and their social spaces: A study of well-being and attachment to place in Aotearoa New Zealand,' *Social Science and Medicine* 68, 664-671.

Willcocks, D.M., Peace, S.M. and Kellaher, L.A. (1987), *Private Lives in Public Places: A Research-based Critique of Residential Life in Local Authority Old People's Homes* (London: Tavistock Publications).

Williams, A. (2002), 'Changing geographies of care: Employing the concept of therapeutic landscapes as a framework in examining home space,' *Social Science and Medicine* 51:1, 141-154.

——— (2006), 'Restructuring home care in the 1990s: Geographical differentiation in Ontario, Canada,' *Health and Place* 12, 228-238.

Williams, A. and Cooper, B. (2008), 'Determining caseloads in the community care of frail older people with chronic illnesses,' *Journal of Clinical Nursing* 17:5A, 60-66.

Wolch, J. and Philo, C. (2000), 'From distributions of deviance to definitions of difference: Past and future mental health geographies,' *Health and Place* 6:3, 137-157.

Wolpert, J. and Wolpert, E. (1976), 'The relocation of released mental hospital patients into residential communities,' *Policy Sciences* 7, 31-51.

Wong, R., Peláez, M., Palloni, A. and Markides, K. (2006), 'Survey data for the study of aging in Latin America and the Caribbean: Selected studies,' *Journal of Aging Health* 18, 157-179.

World Health Organisation (2000), *Community Home-based Care: Family Care-giving – Caring for Family Members with HIV/AIDS and other Chronic Illnesses: The Impact on Older Women and Girls, Botswana Case Study,* (Geneva: WHO).

Wu, P. and Miller, C. (2005), 'Results from a field study: The need for an emotional relationship between the elderly and their assistive technologies,' *Foundations of Augmented Cognition* 11, 889-898.

Yao Yuan (2001), 'Weakening family support for the elderly in China: A cultural explanation,' *Chinese Sociology and Anthropology* 34:1, 26-34.

Young, I.M. (1990), *Justice and the Politics of Difference* (Princeton, NJ: Princeton University Press).

Zarb, G. (1993), 'Ageing with a disability,' in Johnson, J. and Slater, P. (eds) *Ageing and Later Life* (London: Sage).

Zhan, W. (2002), 'China's strategic options for solving the problem of ageing in the 21st century,' *Chinese Sociology and Anthropology* 34:2, 69-74.

Zhang, Y. (2001), 'Survey and reflections on the state of care for the rural elderly in the underdeveloped central region,' *Chinese Sociology and Anthropology* 34:2, 13-23.

Index